WATER HARVESTING, CONSERVATION AND IRRIGATION IN MEWAR

Water Harvesting, Conservation and Irrigation in Mewar (AD 800–1700)

B.L. BHADANI

CENTRE OF ADVANCED STUDY
Department of History
Aligarh Muslim University

MANOHAR
2012

First published 2012

ISBN 978-81-7304-965-1

Published by
Ajay Kumar Jain *for*
Manohar Publishers & Distributors
4753/23 Ansari Road, Daryaganj
New Delhi 110 002

Typeset by
Kohli Print
Delhi 110 051

Printed at
Salasar Imaging Systems
Delhi 110 035

To
the memory of
my mother
Smt. Kesar Devi

Contents

Illustrations

Abbreviations

Ain	*Ain-i-Akbari*
Amarkavyam	Ranchod Bhatt, *Amarkavyam*
EI	*Epigraphia Indica*
Glossary	H.H. Wilson, *A Glossary of Judicial and Revenue Terms*
Khyat	Muhnot Nainsi, *Khyat*
PIHC	*Proceedings of Indian History Congress*
Rajprashasti	Ranchhod Bhatt, *Rajprashasti Mahakavyam*
Raj Vilas	Kavi Maan
RORI	Rajasthan Oriental Research Institute
Tod	James Tod, *Annals and Antiquities of Rajasthan*
Vir Vinod	Shyamal Das

Preface

My interest in the study of water harvesting and water conservation in medieval India goes back to the time I was working for my doctoral thesis on the economic history of Rajasthan. Since there is an intimate relationship between water harvesting and irrigation on which the agriculture of an arid and hilly region as that of Rajasthan heavily depends, I delved deep in water harvesting and conservation for Rajasthan which needed greater and exclusive attention for a detailed understanding of the devices and methods employed in them.

Fortunately, enough material is available in the form of surviving structures, inscriptions, historical works, *farmans*, letters, revenue records, and so on for one to undertake a fruitful study of water harvesting and conservation in Rajasthan. The present study is primarily based on an examination of the extant structures used for water harvesting and conservation purposes in the region of Mewar. The terminal point to include such structures in the study has been the end of the seventeenth century. I have tried to cover major water bodies in the region of Mewar. Yet after an exhaustive physical survey of the surviving structures and archival material, I feel there is much work still left to be accomplished. I can only claim that the study has covered major water harvesting structures that would enable the scholars to comprehend the basic features of the water harvesting system in Rajasthan, particularly Mewar.

At the outset I would like to place on record my gratitude to Professor Ahsan Jan Qaisar for going through the first draft of my chapters on dams. Professor V.S. Bhatnagar from the University of Rajasthan, Jaipur, very kindly agreed to vet the manuscript officially in spite of his health problems. I am indebted to him for his careful vetting and some very useful suggestions.

I am also obliged to my friend Professor V.P. Mittal, Department of Civil Engineering for his valuable advice about the technical aspects of dams and who also helped me with conversion of medieval units of measurements into corresponding modern equivalents.

I am indebted to my friend Professor Ravindra Kumar, IGNOU, New Delhi, for his valuable suggestions. To my friend and colleague Dr. Mohammed Afzal Khan, I owe special thanks for his comments on the technical aspects of the water bodies. Dr. M.K. Pundhir helped me in calculating the measurements of the structures. I am grateful to my friend and colleague Dr. S. Bashir Hasan for rendering help and for going through the entire text.

My student Dr. Jibraeil deserves special appreciation for his help. He made a chart of wells for me for which I express my gratitude. Ms. Safiya Khan showed keen interest in the work. Her constant scrutiny of the subject helped me in many ways.

Ghulam Mujtaba (photographer) and Zameer Ahmad, technical assistant of our department deserve special mention for the photographs and drawings of the water bodies.

My young friend Faiz Habib deserves appreciation for preparing the maps for this book. Fahimuddin deserves thanks for his very efficient and quick word-processing.

My thanks are also due to the staff of the Research Library, Department of History, Aligarh Muslim University, Aligarh.

Finally, my earnest thanks are due to Shri Ramesh Jain, Manohar Publishers & Distributors, for publishing my book and bringing it out in the shortest possible time.

Aligarh B.L. BHADANI

CHAPTER 1

Mewar: History and Geography

The principality ruled over by the Sisodias had acquired the designation Mewar which remained popular in official as well as public mind till Independence. After the foundation of the capital city Udaipur by Rana Udai Singh in *c.*1559, the nomenclature of the state was sometimes alternatively identified with it. Similarly, sometimes Chittor was equated with the state, particularly during the period of its glory before it was captured by the Mughals in the latter part of the sixteenth century. The unique characteristic of the autonomous Udaipur-based Sisodia principality was that its boundaries remained unaltered in pre-Mughal and Mughal times except when it was occupied by the Mughals. But after the treaty of 1615, the boundary of Mewar remained almost unchanged.[1]

Abul Fazl, writing about the territorial expanse of Mewar at the close of the sixteenth century, says that the state is formed by the whole of the *sarkar* of Chittor, *suba* Ajmer. Its territory is large extending about 160 km. in length and about 120 km. in breadth. Fazl further mentions that 'Mewar formerly controlled much more extensive territories, so much so that Rana Sanga possessed a force of 180,000 cavalry and numerous infantry'.[2] The Mughal court historian seems to have recorded the extent of the Sisodia territory when it was under the possession of the Mughals. Therefore, his estimation is probably based on the presumption rather than actual measurement because the entire territory of Mewar never came under the control of the Mughals. The actual area at the close of the sixteenth century could only be worked out by plotting villages of all the *parganas* of the *sarkar* Chittor.

During the fifteenth and till the first quarter of the sixteenth century, the territory of the Sisodia state was vast but suffered reduction thereafter. However, in the late sixteenth and seventeenth centuries it remained uniform and there appears to be no major change in it. The *Vir Vinod* records its length from the north to south as 236.16 km. breadth as 260.86 km. from the east to the west. The total area thus comes to 33,100.77 sq. km.[3] Erskine mentions the area of the state as 32,489 sq. km.[4] Ojha, too, gives identical figures.[5]

The list of the *parganas* under *sarkar* Chittor confirms the limit of the principality of Mewar. But later on, it was split, probably before 1650, into two *sarkars*, i.e. those of Chittor and Kumbhalmer. The former contained 29 *parganas* while the latter 85.[6] Both were the constituents of Mewar. The village list of individual *parganas* under Maharana Raj Singh is available which could give an exact boundary if plotted on the map.[7]

The territory of Mewar has a distinct linguistic identity. Grierson is not right when he classifies it as a part or a sub-dialect of Marwari.[8] He writes that 'further to the south-east in Mewar and the neighbourhood, there is a well-known eastern form of Marwari, known as Mewari or Merawari according to locality'. At another place he records the number of people who spoke Mewari as well as Marwari.[9] In fact, two dialects are spoken in the state, i.e. Mewari and Bagri. The latter is prevalent in the southern and western parts of the region which is the language of the Bhils.[10] In the east, the Kheradi dialect is used which is a mixture of Mewari, Dhundhadi and Hadauti.

I

The development of water harvesting technology is intimately connected with the establishment of the Guhilot rule over Mewar. Therefore, we must trace the history of the state from the ascendancy of the Sisodias in the Aravalli Valley until the time of Maharana Jai Singh (1680–98). The predominant trait of this hilly state was the development of water harvesting structures in the area since the earliest times. The role of the Guhilot rulers in water harvesting and its use thenceforth has been noteworthy.

Guhdatta is said to have been the progenitor of the Guhilot family that ruled over Mewar. However, scholars are not unanimous about the period of his reign.[11] A scrutiny of newly discovered inscriptions reveals the period of this pioneer ruler to be the first quarter of the sixth century.[12] After Guhdatta we simply get the names of his successors but no details of their contribution to the dynasty. Bhoja was the second in line, who is said to have constructed a tank at Eklingji.[13] Next in line were Mahendra and Nagaditya. Siladitya, the son of Nagaditya, emerged as an outstanding ruler in *c.* AD 646. The next prominent ruler was Kalbhoja (AD 712–63) who is identified with Bapa Rawal.[14] He is said to have constructed the temple of Eklingji.[15] His rule between VS 769/AD 712 and VS 820/AD 763–4, suggested by Tod, appears more plausible.[16] He is said to have established his rule over Chittor. Then the name of Allata appears in the contemporary inscriptions who is stated to have shifted his capital to Ahar along the bank of the river of the identical name. After him, many rulers ruled over the region but no significant incident occurred during their reign.

More reliable evidences are available from the time of Rana Ratan Singh who succeeded Rana Samar Singh in 1301–2. He faced the Chittor campaign of Alauddin Khalji in the first decade of the fourteenth century and suffered defeat after putting up a valiant fight. Chittor went into the hands of the Khaljis and then to the Tughlaqs, who retained it for about thirty years.

The territory of Mewar was regained in VS 1394/AD 1337 by Hamir.[17] He ruled over Mewar till 1364 and then his son Kheta succeeded in the same year. His death in 1390 paved the way for his son Maharana Lakha.[18] He got excavated the tanks at Chittor, Kelwara and Lakhawali. His son Mokal ascended the throne between VS 1476/AD 1419 and VS 1477/AD 1420. His assassination in 1433 threw Mewar into political instability.

CONSOLIDATION OF POWER UNDER MAHARANA KUMBHA (1433–1468)

The ascendancy of Maharana Kumbha saw Mewar emerge as one of the most powerful states of India. He suppressed internal dissentions with a heavy hand. He led expeditions into the neighbouring states of Sirohi, Harauti, Dungarpur and others and recaptured territory from them. His emergence also alerted the Muslim states of Malwa and Gujarat. He faced their incursions with full confidence and successfully repulsed them. Thus, his rule in the middle of the fifteenth century was a period of remarkable stability for Mewar. He ruled over a vast territory and consolidated his position, emerging as a rival to the contemporary rulers of Malwa and Gujarat.

Rana Kumbha's reign marks a watershed in the history of building construction, not only in Mewar but in northern India. Besides palaces and temples, his contribution to water harvesting is of immense significance. On his direction, the Ram Kund was excavated with a water-lifting device, probably the Persian wheel (*jalyantra*), besides numerous step wells and tanks at Chittor and other places.[19] Dykes were erected at Kumbhalmer, Vasantpur (in Sirohi) and Abu which demonstrate his interest in the preservation of rain water.

Maharana Kumbha was assassinated in 1468 by his elder son Udai Singh who assumed power and ruled over Mewar till 1473 when he was defeated by his younger brother Rai Mal, whose reign lasted from 1473 to 1509. During his reign Rai Mal remained in conflict with the sultans of Mandu. Other than this his rule remained more or less peaceful and he maintained control over his territorial possessions. He had also erected three tanks namely, Ram, Shankar and Samayasankat.[20]

MEWAR UNDER RANA SANGA (1509–1528)

Mewar became the most powerful state of India in the first two decades of the sixteenth century under its illustrious ruler Maharana Sanga. His contemporary rulers were Sikandar Lodi and Ibrahim Lodi in Delhi, Mahmudshah Begra in Gujarat and Nasirshah Khalji in Malwa. The Rajput rulers of Rajasthan were no match for him. Therefore, he availed this opportunity to expand his territories in different directions. He captured Ajmer, Chatsu and Idar. After this success, he confronted the sultan of Delhi, Ibrahim Lodi. He annexed some territories in eastern Rajasthan that belonged to Delhi, then he defeated the sultan himself at Khatoli. According to Babur, he wrested considerable territory adjacent to Chanderi from Delhi and handed it over to Medini Ray. This success enhanced his prestige in the eyes of his contemporaries. He was successful against Mahmud Khalji II of Malwa whom he kept as a prisoner for some time at Chittor. This shows that the Maharana's power was at its zenith, which is also attested to by the *Baburnama*. At one point of time his ambition was to become the ruler of India but that was dashed in the Battle of Khanwa in March 1526 which he lost to Babur. However, one fact emerges from the statements of Babur—that Rana Sanga had a vast territory yielding 10 crore *dams*,[21] a statement corroborated by Abul Fazl who records that Mewar formerly controlled much more extensive territory.[22]

Sanga's defeat and death was obviously a severe blow to the prestige of the Sisodias but his territorial possessions remained intact under his successors, Ratan Singh and Vikramaditya.

MAHARANA UDAI SINGH (1537–1572)

After a gap of thirteen years, Udai Singh, the fifth son of Maharana Sanga, ascended the throne of Chittor in 1537. After assuming power he first consolidated his position over his territory. Then he led victorious campaigns against Dungarpur and other areas. Though some of his possessions temporarily went into the hands of Sher Shah Suri, this did not affect his position seriously. However, this brought him into conflict with Akbar and led to his defeat at Chittor in 1567. After this Maharana Udai Singh shifted to Udaipur and Kumbhalgarh. Ultimately, the former was chosen as the seat of power but Udai Singh appears to have remained at Gogunda till his death in February 1572.

MAHARANA PRATAP (1572–1597)

Maharana Pratap ascended the throne on 28 February 1572 at Gogunda and after some time a formal coronation ceremony was performed at Kumbhalgarh. He had under his possession only the western hilly tract of Mewar and Godwad; remaining territory, including the forts of Chittor and Mandalgarh, were under Mughals.

Akbar made efforts to enter into a settlement with the Sisodia chief. He sent four political missions between November 1572 and December 1573 under the leaderships of Jalal Khan Korachi, Man Singh, Bhagwant Das and Todar Mal. The failure of the missions resulted in the Battle of Haldighati which was fought in June 1576. After that Mohi and Gogunda passed into the hands of the Mughals. However, the Mughal army, harassed in these areas, ultimately left Mewar and these territories were recaptured by the Maharana.

After the Battle of Haldighati, the tussle between the Mughals and the Sisodia chief over the possession of the hilly areas continued till 1585. The former got success for a while but the area was soon recaptured by the latter. The pressure of Mughal aggression on Mewar eased after 1585 when Akbar decided to concentrate on the Deccan. It is said that during this period Maharana Pratap recaptured major parts of his ancestral territory[23] and made Chawand his centre of power, where he spent twelve years from 1585 to 1597 when he breathed his last on 19 January. Here he built a *kund*, excavated wells and tanks and an embankment for irrigating the surrounding areas. Persian wheels were employed on wells (Figure 1.1).

PEACE TREATY OF 1615 BETWEEN THE MUGHALS AND THE SISODIAS

Amar Singh assumed power after the death of his father Rana Pratap. Mewar was still experiencing political disturbance. The Mughal invasions continued under Prince Salim (1600), Mahabat

Figure 1.1: Persian-wheels in Chawand

Khan (1603), Parwez (1605–6), Abdullah Khan (1609–11) and Khurram (1613–15). The occasional war-like conditions appear to have exhausted both, and this resulted in an honourable agreement between the Sisodia chief Amar Singh and Jahangir in February 1615.[24] Through this treaty, all the territory of Mewar was restored to the Sisodias which is evident from the *firman* issued to Prince Karan, son of Maharana Amar Singh.[25] These are the same *parganas* recorded under *sarkar* Chittor in the *Ain*.[26] These twenty-six *parganas* remained under the possession of successive Sisodia rulers Amar Singh (1597–1620), Karan Singh (1620–8), Jagat Singh (1628–52), Raj Singh (1652–80) and Jai Singh (1680–98).

The period from 1652 to 1698 marks a watershed in the history of water harvesting and conservation system. Both the Maharanas Raj Singh and Jai Singh excelled in the field.

II

The term Mewar appears to have its root in its geographical surroundings. It is encircled by the hills and mountains of the Aravalli ranges. Therefore, there is a possibility that the term is a corrupted form of the Sanskrit word *med pat*,[27] meaning the country of mountains, not 'Meds' and 'Meos' as postulated by some British administrators.

That Mewar is a country of hills and mountains is true to some extent. Roughly two-thirds of the region is covered by plain country, and the remaining one-third is hilly and mountainous. Most of the plain country lies in the northern and eastern parts, which have scattered hills. The south and west present a contrasting picture. These parts are covered with rocks and hills and are called *magra* in local parlance. This portion contains the largest and broadest part of the Aravalli mountains (Figure 1.2).

The western part is dominated by the Aravalli hills that enter from Merwara. At this point its height is 727 m. above sea level. Then the hills extend in the south-westerly direction and their height reaches up to 1,088 m. at Kumbhalgarh. Their height

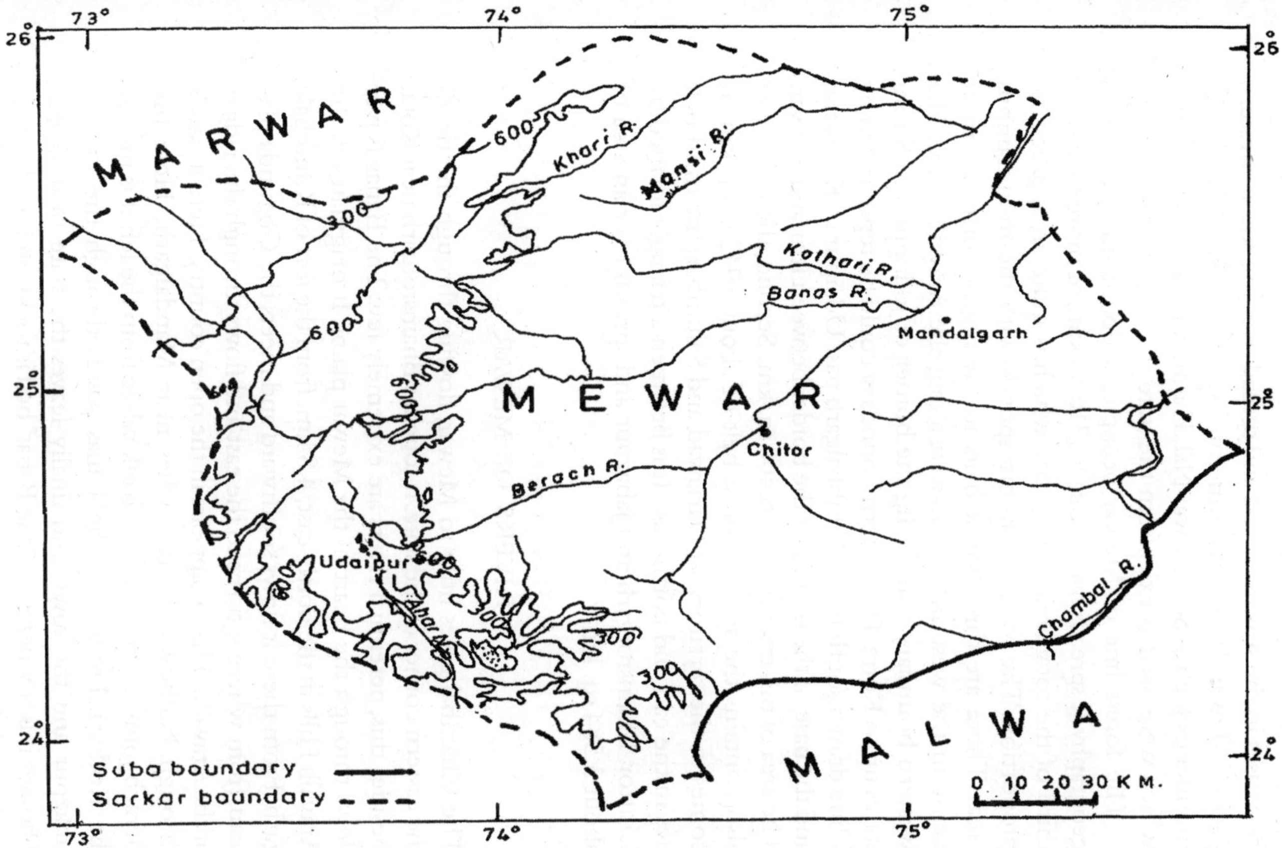

Figure 1.2: Map Showing Geography of Mewar

decreases in the southern part and spread over the south-western part of Mewar. There are numerous passes, known as *nal*, in the mountains such as Someshwar Nal, Hathidara Nal and Paglia Nal, which were used as exits from the state.[28]

The forest line of Mewar occupies a dominant place in its geography. It spreads over about 11,716 sq. km., or more than one-third of the total area of the state which is quite an appreciable proportion. The forest area corresponds with the mountainous areas. There are three forest belts in the region: first, the most dense in the west and south-west along the Aravalli range. The western boundary runs along the border of Jodhpur and Sirohi states up to Kotra. The eastern boundary could be represented by a line drawn south from Kumbhalgarh past Udaipur to Khewara; and the base of the triangle is the border between Idar and Mewar. The area of this tract is about 6,400 sq. km. Second, the south-east also contains the second forest belt of about 1945.6 sq. km. It comes in the territory of Dariawad and Salumbar; the third belt lies in the east and north-east. It is between a triangle formed by Chittor, Bhainsrorgarh and Jahazpur and spreads over an area of about 3,584 sq. km.[29]

RIVERS OF MEWAR

The Chambal gives little to Mewar, flowing through the tip of the eastern corner of the state. It touches Bhainsrorgarh near Kota. Besides this, perennial rivers are extremely few. The Banas River flows through the heart of the Mewar plains. It originates in the Aravalli Hills in the north-east, 4.8 km. from the fort of Kumbhalgarh. From there it flows southward and meets the Gogunda plateau, from where it turns to the east and flows through the ridges of the Aravalli. Then it appears in the open country. Here it passes through Nathdwara and reaches near Mandalgarh. Here the Berach joins it from the right bank while from the left it is met by the Kothari. From here the Banas passes through the hills of Jahazpur and the town and finally leaves the state near Deoli. Though it is not perennial the Banas' bed is rocky and hard which benefits Mewar. Since the water is retained under the surface for a longer period. Wells were dug on both the banks to tap the underground water. The biggest beneficiaries of this were those villages which were settled along the river.

The Khari emerges from the hills of Diwer and passing the north-east of Deogarh enters Ajmer. Finally, it merges into the Banas in the territory of Jaipur.

The river Berach, also called Ahar, originates in the hills to the north of Udaipur. Its course is very short. From Udaipur it passes through Bedla and falls into the Udai Sagar Lake, constructed by Maharana Udai Singh. From there it is called Udai Sagar-ka-Nala. It goes further to the east and reaches Chittor, where it turns north-east and merges into the Banas, near Mandalgarh.

The Koteshari, popularly known as Kothari, flows to the south of the Khari. It emerges from the Aravalli and passes south of Diwer and then merges into the Banas at a distance of about 4 km. from Nandrai.

The river Wakal emerges from the hills west of Gogunda. It flows between Oghana and Manpur located to the south. Then after taking a north-west turn reaches Kotra. Finally, it takes a turn to the west and merges into the Sabarmati in the territory of Idar.

Another river is Som which receives the drainage of the mountains in the south-western direction and water from the outlet of the Jai Samand Lake. First it goes to the boundary of Dungarpur to join the Jakam, then turns near village Babrana and falls into the Mahi.

The river Jakam emerges from the hills near Chhoti Sadri and goes south and enters Pratapgarh where the river Karmari merges with it. Then it re-enters Mewar, runs in the south-westerly direction near Dhariyawad and finally merges into the Som.

The above-mentioned eight rivers made the territory of Mewar fertile by raising the level of groundwater along their banks. Ordinarily, water could be found near the surface of the banks of these rivers. This is the reason that majority of the wells were dug out around the beds of these rivers. To tap this underground spring, wells were dug on both sides of beds of rivers. These kinds of water monuments were known as the *seja* wells (shallow or spring wells) in local parlance whose depth ranged between 7.5 and 9.0 m. While another kind of well was designated as the *akhara* which was the deeper 12.2–15.25 m. than the *seja*. The *akhara* kinds of wells are found in large numbers in Mewar.

Generally, two types of water-lifting devices appear to have been in operation: *charas* (leather-bucket) and *rahant* (Persian wheel). The former apparatus was employed on the wells of the *parganas* located in the east and north while the latter machine was used in the areas of the south and the western part of the state. This means that the water level was nearer to the surface in the east and north of the state, while the south and the western part contained water at considerable depth. This largely corresponds with the physiography of the area. The northern and eastern portions broadly comprise open country and the plains with some scattered hills here and there, while the southern and western parts are covered with rocks, hills and dense jungles. This area was styled as the hilly tracts of Mewar by the British administrators.

Shyamal Das, the author of *Vir Vinod*, raises a question about well and canal irrigation. He writes that there was no system of agricultural irrigation before 200 years ago. That is why except the wells for drinking water and for irrigating the gardens there were no signs to show that agricultural fields were also irrigated in the past. No canals were even dug out from the tanks. Das' statement is not based on facts because it is negated by the extant evidence.

The geographical description of Nainsi too is quite informative. According to him, Mewar was largely formed of valleys of rivers flowing eastwards from the Aravalli Hills. On the western side, the *Khyat* records, were the mountains in the *parganas* of Jhilwara, Kelwa, Kumbhalmer, Desuri, Gogunda and Jharol. It can be called the western mountain chain, extending from Jhilwara to Jharol.[30] On the southern side there was a noted hill range extending from Sadri to Dungarpur and Banswara.[31]

The geographical and physiographic conditions of Mewar have naturally affected the pattern of rural settlements in the region. The northern and eastern sides of Mewar consist of an elevated plateau. Here the *Khyat* mentions *pathar* (plateau) villages. There was a cluster of villages on the mountain of the Akhan near Chittor.[32] In the tract between Chittor and Mandalgarh–Bhainsrorgarh, the rural settlement was appreciably dense, more than 300 villages were spread over this part of the state.[33] The area of Nahesar was remarkably fertile and, therefore, there were as many as 900 villages. Each village possessed 500 to 600 *bighas* of cultivable land.[34] Similarly, the Merwara and Mandal districts comprised 140 villages each.[35]

From Nainsi's account the relationship between the mountain, its water resources and the rural settlement emerges very prominently. Whenever Nainsi gives a description of the mountains and water resources, he purposely goes on to give the number of the rural settlements. From the *Khyat* it becomes clear that many mountains were rich in water resources that attracted settlement. For example, the Machhawala mountain and the hills of Rahag possessed abundance of water.[36] Consequently, these areas witnessed a high density of rural settlement.

We can compare Nainsi's information about rural settlements with the population map in the *National Atlas of India*[37] drawn on the basis of the 1951 census. The rural settlements in the south-west were fewer and widely scattered compared to the north-east. Indeed, we can say that the general pattern of rural settlement in the *Khyat* broadly coincided with the pattern of the density of population in 1951.

III

It is believed that the Guhils first established their seat of power at Nagda,[38] 22 km. north of Udaipur and about 1 km. west of Eklingji. The Guhil chief appears to have selected this site because of an enormous natural waterbody which could serve a large number of people and irrigate a large area (Figure 1.3). This structure still exists but its name has changed to Baghela or Baghelao Talab.[39] It was constructed by Rana Mokal (1419–33) in memory of his issueless brother Rawat Bagh[40] (Figure 1.4). After Nagda, Ahad got the status of capital, probably at the end of the tenth century

Figure 1.3: Natural Depression at Nagda

Figure 1.4: Baghela Talab at the Site of Depression at Nagda

though the area was already under the Guhil rulers.[41] The most likely reason for shifting the capital was its nearness to a river or stream of the same name. The capital is about 3 km. east of the city of Udaipur and was founded in the sixteenth century. Ahad's location on the trade junction could be another reason for shifting the capital.[42]

It is said that wells were excavated there by the Guhil rulers too. This can be deduced from an inscription of AD 1179, which records the grant of *arghata* (garland of pots) with a plot of cultivable land.[43]

Next Chittor got the status of capital in the first half of the thirteenth century,[44] having served as the royal seat or seat of government of many powerful dynasties in the past such as the Hunas, Moris, Gurjara–Pratiharas and the Parmars of Malwa. The area also remained under the possession of the Chauhans. This is evident from the Bijolia inscription of 1168 and the Menal inscription of 1170.[45] The fort of Chittor, after its occupation by the Guhils of Nagda–Ahar region, became their capital. Chittor was the place which benefited the most under the illustrious rulers of these powerful dynasties. Numerous water bodies were excavated and constructed by them in the fort which could help its residents survive for longer periods during times of hostilities.

The next capital, Udaipur, was planned by Maharana Udai Singh visualizing the vulnerability of the Chittor Fort. Nainsi writes that about seven years prior to the fall of the great citadel, the Sisodia chief constructed the Udai Sagar Dam and founded Udaipur.[46] Udai Singh is said to have never come to the newly-founded city, he remained at Gogunda and breathed his last there.[47]

The site for the new city was selected after long deliberations with nobles and civil engineers. The availability of water in and around the area appears to have been given foremost priority while selecting the site. It seems that the existence of various sources of water coming from the surrounding mountains attracted the attention of the Maharana and for this reason the Pichhola Lake was chosen as the residential complex of the royal family and the open area was left for the common people.[48]

To sustain the large population of the prospective city, a water body was constructed to the east of Udaipur and named after its founder, viz., Udai Sagar. This water body was to receive water from various streams.[49] This shows that the availability of water was the main concern of the Sisodia chiefs for the selection of the site for their capital. They also utilized ingenious techniques for raising water harvesting structures in the area.

Udaipur served as the capital of Mewar till 1947 except for a short period of the Mughal occupation from 1567–1615. During this unwelcome interval Gogunda enjoyed the status of being the royal seat and benefitted from the construction of water bodies by the son of Udai Singh, Maharana Pratap. He had chosen this site for his shelter as well as for royal seat because of abundant water in the area.

The above discussion shows that the geography and topography of Mewar offered many largesse to the Sisodia rulers such as the range of mountains which served as a formidable barrier against aggression, abundant water sources, vast areas under forest, numerous seasonal rivers emanating from the surrounding hills, natural depressions and the open plains suitable for agriculture. The rulers seem to have made plans keeping in mind the extent of the territory under their possession. They wisely utilized nature's gifts on two fronts: selection of site adjacent to natural water source for the royal seat and the development of well-knit water harvesting and conservation structures tapping both surface as well as underground water. The system of distribution of the accumulated water, too, was developed by constructing canals for irrigating vast agricultural tracts. Abundance of water led them to plan and design gardens and orchards which also contributed in creating a pleasant environment. The efforts of the Sisodia rulers in the field of water harvesting and conservation system and its distribution is the theme to be investigated in this monograph.

NOTES

1. In modern times four districts, namely Udaipur, Bhilwara, Chittor and Raj Nagar constituted Mewar in the pre-1947 period.
2. Abul Fazl, *Ain-i Akbari,* vol. II, tr. H.S. Jarret and 2nd edition corrected and annotated by Jadunath Sarkar, Calcutta: Orient Book Corporation, 3rd edn., 1978, pp. 273–6.

   The Hindi word *kos* is derived from the Sanskrit *krosa* which is equivalent to 4 km.
3. Shyamal Das, *Vir Vinod,* I, Delhi: B.R. Publishing Corporation, rpt. 1986, p. 100.
4. K.D. Erskine, *Rajputana Gazetteers: The Mewar Residency*, vol. I, Gurgaon: Vintage Books, rpt. 1992, p. 5.
5. G.H. Ojha, *Udaipur Rajya ka Itihas*, vol. I, Jodhpur: Rajasthani Granthagar, rpt., 1999, p. 2.
6. *Suban-ri-Sirkar-re-Pargana-ri-Vigat* (Ms), No. 227/3, Anup Sanskrit Library, Bikaner. This manuscript records 85 *parganas* under the *sarkar* Kumbhalmer while in the printed text of Muhnot Nainsi the total is entered as 95 which appears either a printing error or a slip in the original MS because the breakup given below comes to 85 (cf. *Khyat*, I, ed. B.P. Sakariya, Jodhpur: Rajasthan Oriental Research Institute, 1960, p. 53). The evidence of Nainsi for the *sarkars* of Chittor and Kumbhalmer has been cited by Irfan Habib in his *An Atlas of the Mughal Empire*, sheet 6A. Delhi: Oxford University Press, 1982; however, the page nos. are 52–3 and not 42–3 as he cites. The total of the *parganas* under the *sarkar* Chittor of Nainsi does not match with the *Suban-ri-Sirkar-re-Pargana-ri-Vigat*. In the former the number given is 27 while 29 is mentioned in the latter work.
7. *Patta Bahi Pattedaran-ri-Vigat*, I, and *Maharana Raj Singh Pargana Bahi*, II, ed. Hukum Singh Bhati, Udaipur: Himanshu Publications, 1995.
8. G.A. Grierson, *Linguistic Survey of India*, vol. IX, part II, Delhi: Low Price Publications, rpt. 1990, p. 16.
9. Grierson incorporates more areas where Mewari is spoken such as Tonk, Pratapgarh, Ajmer, Merwara, Kishangarh, besides Mewar. He gives the total number of persons as 1,684,864, who spoke this dialect (cf. ibid., pp. 16–17 and 76).
10. K.D. Erskine writes that more than 90 per cent of the people

spoke Mewari and 6 per cent used Bhili or Vagdi. But he too identifies it as a variety of Marwari, cf. *Rajputana Gazetteers*, op. cit., p. 35. But G.H. Ojha identifies Mewari as a separate dialect, op. cit., pp. 13–14; Grierson too records a separate dialect of the Bhils which he styles as *Magra-ki-Boli*, op. cit., p. 70.

11. Ojha assigns VS 623/AD 566 as the period of Guhadatta's reign. He deduces this date from an inscription of VS 703/AD 646. There is a reference in it about Siladitya who has been described as a great grandson of Guhil. He assigns 20 years to each and thus gets this date by deducting 80 years for the four generations from the year VS 703/AD 646 (cf. op. cit., p. 98).
12. Ram Vallabh Somani on the basis of numerous inscriptions and the examination of genealogies of Nagar, Vagad and Mewar assigns later than first quarter of the sixth century to Guhadatta. See, *History of Mewar* (*From Earliest Times to 1751 AD*), Jaipur: C.L. Ranka & Co., 1976, pp. 31–5.
13. A.K. Vyas, 'Kadmal Plates of Guhila Vijayasimha, V.S. 1140', *Epigraphia Indica*, XXXI, 1955–6, ed. D.C. Sircar, Delhi: Department of Archaeology, Calcutta, 1960, pp. 237–48.
14. Ojha, op. cit., pp. 102–4; Somani, op. cit., p. 40. There are divergent views about his identity. For details, sec Somani, op. cit., pp. 40–3.
15. A.K. Vyas, 'First and Third Slabs of Kumbalgarh Inscription VS 1517', *Epigraphia India*, vol. XXIV, 1937–8, ed. N.P. Chakravarti, Delhi: Manager of Publishing, 1942, pp. 304–28.
16. James Tod, *Annals and Antiquities of Rajasthan*, I, London: Routledge and Kegan Paul Ltd., rpt., 1960, pp. 182–9. For a different view, see Somani, op. cit., pp. 43–4.
17. In VS 1393/AD 1336 Hamir occupied Chittor, cf. 'Mewar Rawal Ranaji-ri-bat' *Majjhmika*, vol. 17, Udaipur: Pratap Shodh Sansthan, 1993, p. 13. But Somani assigns the year of occupation as VS 1394/ AD 1337, cf., op. cit., p. 107.
18. Ojha assigns AD 1382 as the death of Kheta, cf. op. cit., p. 256 but Somani differs on this point, cf. ibid., p. 111.
19. Ojha, op. cit., I, pp. 309–10.
20. Nainsi, op. cit., p. 17; Ojha, op. cit., pp. 3, 22, 26, 44.
21. Zahiruddin Muhammad Babur, *Baburnama,* tr. Annette Susanah Beveridge I & II, Delhi: Oriental Books Reprint Corporation, rpt., 1979, pp. 483, 562.

22. Abul Fazl, op. cit., p. 276.
23. Somani draws this conclusion after interpreting inscriptions; op. cit., p. 240.
24. Ojha, op. cit, pp. 496–7; Somani, op. cit., p. 257.
25. The list of *parganas* given in the *farman* are reproduced in *Vir Vinod,* II, part I, Delhi: B.R. Publishing Corporation, rpt. 1986, pp. 239–49; Ojha, op. cit., pp. 502–3.
26. *Ain*, II, pp. 278–9.
27. The term *Medpat* was employed as synonymous for Mewar. A Jain poet named Hem composed a bilingual poem entitled 'Medpat Varnana Prashasti' in Sanskrit and Rajasthani narrates the characteristics of Mewar. Similarly, in another poem the same poet used Mewar for the Sisodia territory which makes it clear that both the terms were interchangeably employed cf., 'Medpat Varnana Prasashti' and 'Mewar Desh Chhand', in *Maru Bharti*, July 1954, II, part I, Pilani: Birla Education Trust, 1954, pp. 49–50.
28. K.D. Erskine, ibid., pp. 5–6; V.C. Mishra, *Geography of Rajasthan*, Delhi: National Book Trust, 1967, pp. 33–4.
29. The description about the forest is entirely based on Erskine, op. cit., p. 51.
30. *Khyat,* I, pp. 40–3; Erskine, op. cit., p. 6.
31. Erskine, op. cit., p. 6.
32. *Khyat,* I, pp. 43-4.
33. Ibid., p. 44.
34. Ibid., p. 46.
35. Ibid., p. 45.
36. Ibid., p. 41.
37. *National Atlas of India*, Population Jaipur Plate, Survey of India Maps and Census of India 1951, Government of India, 1960.
38. Ojha, op. cit., p. 98; K.C. Jain, *Ancient Cities and Towns of Rajasthan: A Study of Culture and Civilization*, Delhi: Motilal Banarsidass, 1972, pp. 213–19; Nandini Sinha Kapur cites the wrong page number (fn.19) of K.C. Jain's book, *State Formation in Rajasthan*, Delhi: Manohar, 2002, p. 35.
39. Ranchhod Bhatt, *Rajprashasti Mahakavyam*, ed. M.L. Menaria, Udaipur: Sahitya Sansthan, Rajasthan Vidyapeeth, 1973, pp. 39–40. We have surveyed this structure and undertaken extensive measurement which demonstrates its enormity. Rana Mokal simply constructed an embankment wall and built *ghat*, stairs and *chhatris.*

Its extent is as vast as it was in the seventh century. This shows the wisdom of the chief behind the selection of the first seat of power.

40. *Rajprashashti*, op. cit., p. 40.
41. Jain, op. cit., p. 220; Nandini Sinha Kapur, op. cit., pp. 57–8.
42. Ibid.
43. The inscription of AD 120 is available in the *Annual Report Rajputana Museums,* Ajmer, 1931, p. 4 as quoted in Jain, op. cit., p. 222. Though the grant was given by Bhimdeva II, the Chalukya ruler of Gujarat to a Brahman, Ahar came under their possession for a short period which indicates the construction of wells by the Guhils in the region and not by the Chalukya ruler.
44. There is disagreement over the name of the ruler who made Chittor as the capital. Jain assigns its credit to Jaitrasimha (AD 1213–52), op. cit., p. 227, while Nandini Sinha Kapur doubts it and gives credit to Tej Sinha basing her assertion on the inscriptions found at Chittor (op. cit., p. 62).
45. A.K. Vyas, 'Bijoli Rock Inscription of Chahmana Samesvara: V.S. 1226', *Epigraphia Indica,* vol. XXVI, Calcutta: Department of Archaeology, 1952, pp. 84–112. Menal Inscription, *Progress Report of Archaeological Survey, Western Circle, 1905–6*, pp. 57, 60. This can be deduced from the grants made to the temples of Mahavira and Siva at Bijolia and Menal ignoring the Guhils, as quoted by Nandini Sinha Kapur, op. cit., p. 63 and fn. 71.
46. A *charan*, named Khidiya, informed Nainsi that five or seven years prior to the fall of the Chittor, Rana founded Udaipur and that the Udai Sagar Dam was constructed even before that (Nainsi, op. cit., p. 48).
47. Ibid.
48. The descriptions of the planning for the new city and the fort is recorded in detail in Manohar Singh Ranawat (ed.), *Chittor-Udaipur Ka Patnama*, vol. II, Udaipur: Chirag Prakashan, 2003, pp. 230–42; Vyas (ed.), 'Bijoli Rock Inscription of Chahamana Samesvara: V.S. 1226', op. cit.
49. B.L. Bhadani, 'Udai Sagar: A Sixteenth-Century Dam in Rajasthan' in *Medieval India 2*, ed. Shahabuddin Iraqi, Delhi: Manohar, 2008, pp. 273–86.

PART I

WATER HARVESTING AND DISTRIBUTION: DAMS

CHAPTER 2

Indra Sarovar Dam

The tradition of water conservation is as ancient as the Indian civilization. This is witnessed by the innumerable extant structures of pre-colonial period in different parts of the country. However, very few of them have been studied by historians and archaeologists.[1] A majority of them still await detailed examination based on physical survey to understand the development of indigenous technology and its impact on agriculture and ecology.

Undoubtedly, the role of geography and ecology of the Aravalli range in the Mewar region has been as crucial in the building of dams, tanks, wells and step wells as the value system of the Sisodia rulers. In the absence of a perennial river system, it was prudent of the masters of the region to search for other alternatives to make water available throughout the year for drinking as well as for irrigation. Their endeavours had two objectives: first, to achieve the mythological status of Bhagirath, who is said to have brought the Ganga to the earth; and second, the more earthly utilitarian purpose as mentioned earlier. The latter was linked with the economic necessity of the state.

Water harvesting is generally intended for supplying drinking water as well as for irrigation, but sometimes it is undertaken for religious purposes too.[2] Endeavours were made to create artificial structures—*baoris*, tanks, *kunds* and dykes—which could retain water for long periods. These reservoirs catered severally to secular and religious structures such as temples. However, some were erected mainly for religious purposes[3] though with the passage of time they were also used for secular function, that is, irrigation. In general, the sources of water such as rivers, springs and tanks have always had some degree of holiness attached to them. They have been associated with the local pantheon of Hindu gods and goddesses,[4] yet religious reservoirs have been exclusive in nature. A water structure associated with a specific temple or shrine carries its own significance, where a dip in its waters before or after the *darshan* (visit) of the deity is considered sacred. The people assign holy status to it and in course of time, it becomes a centre of pilgrimage. Under this category, a lake of religious sanctity was created at village Kailashpuri, near Udaipur, attached to the temple of Eklingji—the family deity of the rulers of Mewar.[5]

This weir, known as Indra Sarovar, found behind the Eklingji temple,[6] was built by constructing a wall between two hills. This masonry work may be small in size but it perhaps the oldest in Rajasthan and has been totally ignored by historians and archaeologists.[7] An attempt has been made in this chapter to study the different aspects of this dyke based on fieldwork,[8] such as its geographical ambience, date or year of construciton and size.

The village of Kailashpuri is located in a valley about 16 km. north of Udaipur. The old capital of the Guhilots, Nagads, is about 1 km. north of this place. This hamlet was overshadowed by the temple of Eklingji.[9] The temple stands in a narrow triangular gorge formed by the Trikuta[10] range of hills, popularly known as the 'wildest portion of the Aravalli'[11] and surrounded by 'towering hills of primitive formation'.[12]

The rugged and rocky Aravallis (726.82 m.) make their entrance into Mewar from the Ajmer–Merwara side and extend towards the south passing through Kumbhalgarh (1,088.24 m.) and Gogunda (1,316 m.) Eklingji is situated about 20 km. east of Gogunda. The Aravallis then deviate towards the east and are full of seasonal springs[13] (Figure 2.1).

The region of Mewar is devoid of any perennial rivers. The major rivers of the state originate from the north-western and southern mountains. The Banas, the largest river of the region, which rises west of Kumbhalgarh (which also houses a famous fort built by Rana Kumbha in 1458),[14] flows through Gogunda to the west of Eklingji and then turns towards Nathdwara[15] in the north. In the western hills near Gogunda originates another river, Wakal, whose waters overflow and reach Eklingji.

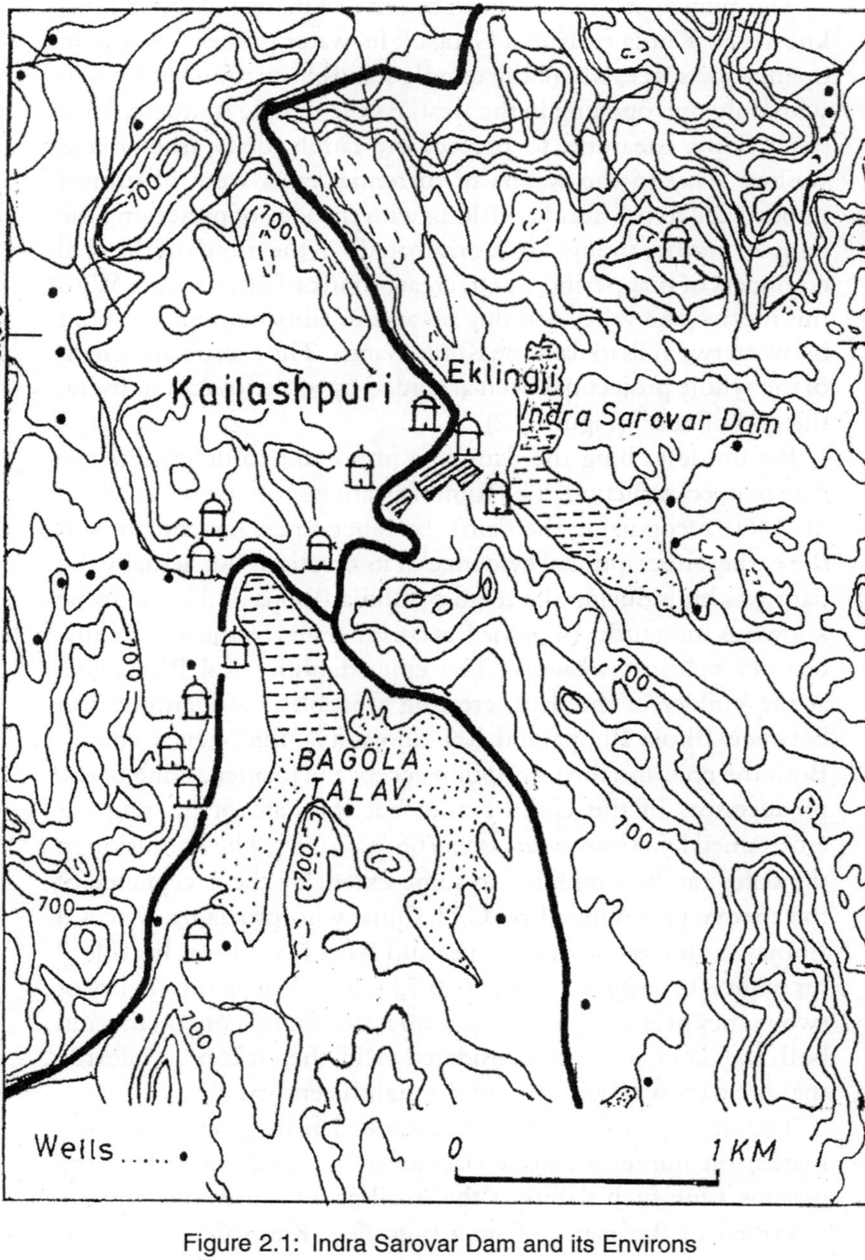

Figure 2.1: Indra Sarovar Dam and its Environs

The rainy season sees the birth of a rivulet in the hills of Trikuta, known as Kutila or Kutila Ganga.[16] Its water gushes towards an open, wide space, where the temple stands,[17] surrounded by hills with only one opening in the west. As the temple is situated on a steep slope, measures to protect the family deity from being washed away in the waters of torrential rains were of utmost importance to the Sisodias. Along with the idea of preserving the temple, another thought that originated[18] in the minds of the civil engineers of that period was to create a tank of sacred value. With this motive, the valley was dug up and an embankment was built between two hills to dam the Kutila Ganga. The success and credit of the whole project has been accorded by some sources to Indra, the god of rain[19] (Figure 2.2).

Before describing the dam, it is important to determine the date of its construction. The Kumbhalgarh inscription of VS 1517/AD 1460, recognizing the dam's existence since ancient times, is the earliest inscriptional evidence of its erection. But actually the dam was built during the reign of Bhoja Bhopa and came to be known as Indratirtha Bhojasar,[20] as corroborated by the sevententh-century epic, *Amarkavyam*. The epic identifies Kal Bhojaditya as the builder of the dam, creating a problem of identification between Bhoja Bhopa and Kal Bhojaditya, the Mewar ruler.[21] Both the epic and the inscription accept the former as one of the predecessors of the Guhil family, but the date of his reign as ascertained by *Amarkavyam* is VS 766/AD 709.[22] The confusion in the dates can be simplified to some extent by the acceptance of the theory propounded by G.H. Ojha, who proclaims that Kal Bhoja was the name of the ruler and Bapa Rawal was his title;[23] his reign is broadly accepted as AD 734–53.[24] If at all this theory of two names personifying one person is not found convincing and both the dates are to be considered valid, then it can be inferred that the dam was built during the eighth century.

The antiquity of the dam can be assessed by the graffitis (masons' marks) on numerous stone slabs over the dam, incised by the masons. One such slab near the small gate towards the south is inscribed in the name of *sutra* (*sutradhar*) Kanji Chatroji[25] and

Figure 2.2: Front View of the Embankment

dated VS 1043/AD 986. Furthermore, the year VS 1167/AD 1107 is carved on another unsigned slab. The engraving on yet another slab reads, 'When the *pal* (embankment) was built in VS 1202/ AD 1145 [?] then stairs [were] added. . .'[26] An outlet on the north facing the reservoir has a stone slab, which though mutilated, shows the date deciphered tentatively as, '*Samvat* 1437'.[27] These different dates presumably relate to the repair work taken up occasionally, but confirm the antiquity of the structure (Figure 2.3).

Masons' marks on any structure are of extreme importance. This particular dam has numerous engravings indicating the involvement of different groups of masons. Besides them, even the name of the engineer is recorded. These records may well be used for comparison with the marks on other buildings in the same locality to confirm the date of their construction (Figure 2.4).

The *Rajprashasti*[28] too confirms the antiquity of the dyke, also documenting the fact that extensive repair work was carried out in VS 1729/AD 1672 by Maharana Raj Singh. It mentions that since the *sopan* (stairs) and *setu* (dam) were tottering, the Maharana got these repaired and also constructed four *pratolias* (platforms) and a strong wall at the cost of Rs. 18,000.[29] According to the Kumbhalgarh epigraph, it was also patronized by the Mewar Maharanas.

The existing structure on the Indra Sarovar indicates that the main wall from the north to the south is not in a straight line and hence, does not connect the hill on the north, but instead tilts to join a *kharanja* (passageway). For this, a 10.3-m. long void was filled with earth and rubble and the dyke was left out to be used as a road to reach the other side of the lake. A small arched bridge (still in existence) below takes the water away to the reservoir on the other side. The primary purpose of this foot-bridge or overpass was to provide water to the agricultural terraces located at the foothills (Figure 2.5). The small structure on the east was used for lifting water from the main tank for irrigation. On the southern side, a wide platform with temple complexes and an open space for the devotees is visible. At this juncture, where the southern

Figure 2.3: Inscription on the Outlet facing the Reservoir

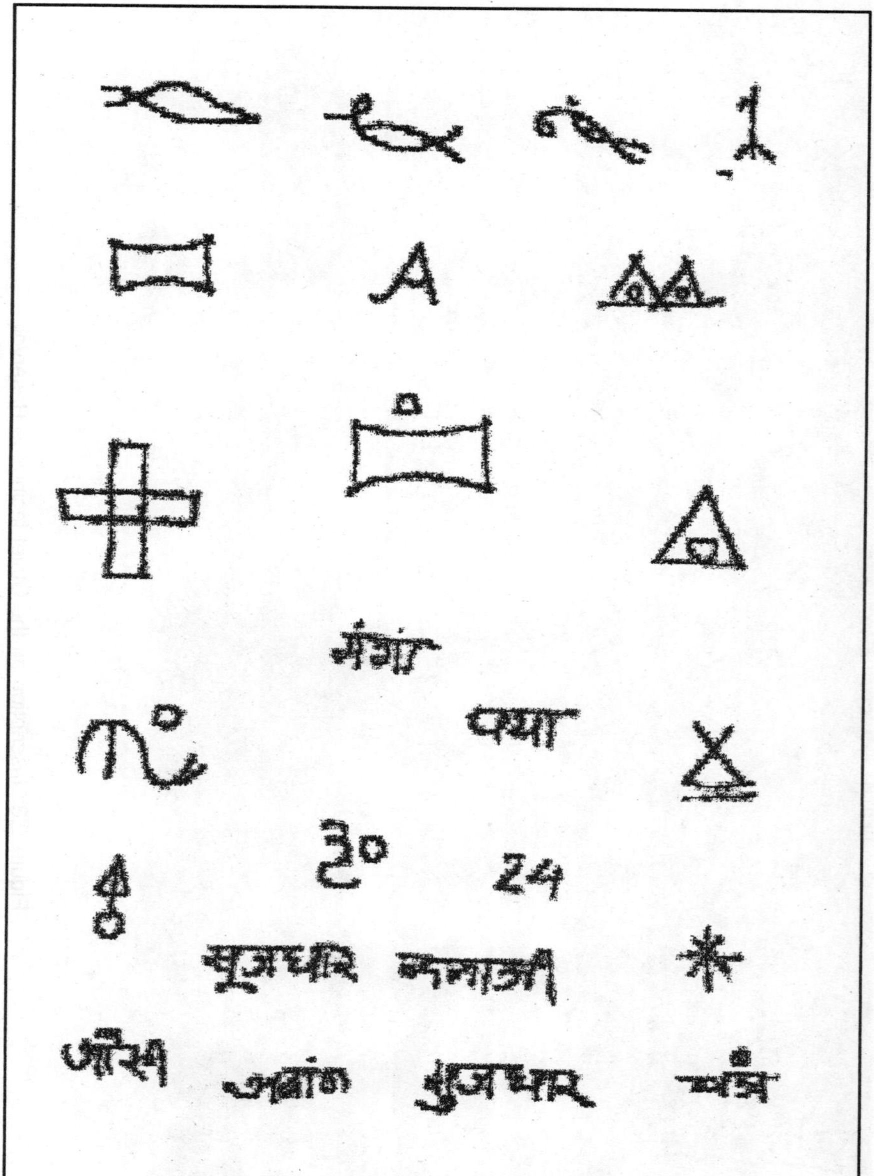

Figure 2.4: Mason's Marks and their Names on Indra Sarovar

Figure 2.5: View of the *Kharanja* (Passageway) and Bridge Bifurcating the Reservoir

wall projects towards the east and west, space was left to erect the temple. To accommodate these, a 30-m. long wall from the south to the north which pushes back and breaks the alignment with the remaining part, was constructed though it disturbed the harmonious lake-side projection (Figure 2.6).

The wall projecting from the plinth towards the lake is in the form of steps, which are 11 in number, each measuring 0.25 m., thus accounting for the 2.75-m. projection. At the end are seven platforms 1 × 1 m in dimension. In between the two estrades are rows of four stairs, the workout width coming to 2.75 m. To the back of the barrage, two walls of 25.5 m. width each, filled with earth and rubble, were erected at the plinth level to give additional strength, thus increasing the total width to 28.75 m. (2.75 m. projection towards the lake plus 25.5 m. of the backwall). About 3 m. of the wall was dressed with stone slabs, which seems to be a later addition.

The height of the dam from the ground to the plinth is 5.25 m. The wall goes 2 m. beyond this level, making the total elevation 7.25 m. There was difficulty in working out the length of the embankment as a portion of it has been blocked with a brick-wall built by the temple authorities. Anyhow, this stretch should be approximately 30 m. in length. The remaining portion was measured in two parts, the straight one measuring 68.50 m., while the diverted one comes up to 38.50 m. The total length, thus, comes to 137 m. (Figure 2.7 layout drawing).

The wall above the plinth level has three gates of different sizes, the main entrance is of 2 m. and other two of 1 m. each. There seems to have been a driving need for this. Firstly, there would be pilgrims keen on taking a holy bath in the tank, and secondly to function as an outlet if an emergency arose due to torrential rains. At the back of the barrage, a sloping passage was made to take away the surplus water on the embankment. This testifies the awareness of the civil engineers concerning the safety of the structure.

A little above the wall, over the plinth, in the southern direction a square opening for the spillway, presently covered, can be seen.

Figure 2.6: Structure for Lifting Water for Irrigating Fields

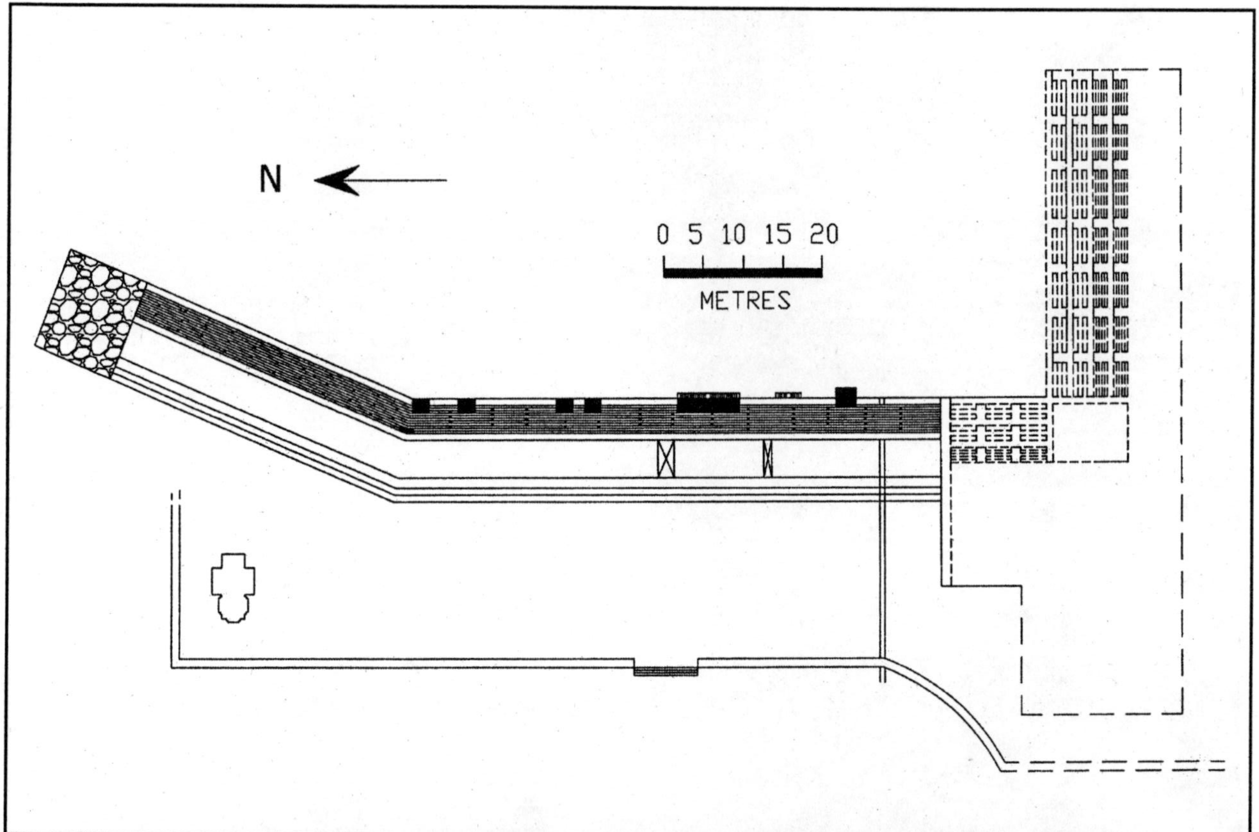

Figure 2.7: Ground Plan of the Indra Sarovar Dam

It was probably constructed on *Sawam Sudi* 11, VS 1767/AD 1710.[30] This was meant to take the surplus water of the tank through the canal to distant areas. The water through the channel first enters a small garden[31] to its west, then comes out and falls into a wider canal along the back wall of the Eklingji Temple, flowing down along the hills to distant localities, though the course is yet to be surveyed (Figure 2.8).

Considering the period of construction of the dyke, one can conclude that it was considerably large in size. Now, the question can be which segment was actually part of the original plan and which sections were added later on? We need to analyse the evidence scattered over the dyke and preserved in records to arrive at some conclusion. First, the inscriptions of different dates left by masons on the body of the structure indicate the various eras of addition and repair. Second, the information yielded by the epigraphs as well as historical epics like the *Amarkavyam* and the *Rajprashasti* also attest to this fact. According to the *Amarkavyam* some repair work was commissioned by Maharana Raj Singh in VS 1729/AD 1672.[32] The date is in consonance with the architectural style of the Raj Samand Dam situated nearby. However, less than one-third of the portion of the Indra Sarovar Dam on the south is not in harmony with the rest of the structure. As a result, it can be considered as a later addition and totally revamped from the foundation. As discussed earlier, the southern side-wall is pushed back and is not in alignment with the whole, thus implying the laying of a fresh foundation. This was done to widen the platform so as to accommodate a group of temples. Similarly, the wall with three main entrance gates from the plinth appears to have been constructed *de novo*.

We surmise that this structure is datable to the eighth century and witnessed major alterations. These modifications seem to have been carried out at two levels: the changes in the south by projecting the platform in the east and west, and the construction of the wall from the plinth level. The base of the wall from the south to the north appears to have been retained, and a new foundation was probably laid to push the platform to the back.

Figure 2.8: View of Outlet: Water Falls into the Canal along the Back Wall of the Eklingji Temple

As for irrigation works on the east and an aqueduct at the back of the structure, it can be easily guessed that the former was in existence since the inception of this establishment. From this, it can be inferred that the entire embankment, except for a segment in the south and a wall from the plinth, including the foot bridge, was in the original plan.

The study of the Indra Sarovar Dam enables us to draw the following conclusions: it was a masonry structure with sufficient breadth, which increased its weight enormously. This is the main characteristic of gravity dams, which ensures protection of the barrier against the pressure of water. Second, the site selection by the architects was guided by two factors: religious and hydrological. The barrage was erected primarily to protect the temple. Thus hydrological factor as well as two other reasons for impounding the water have come to the fore: drinking and irrigation. This is confirmed by the presence of a conduit in the wall and a water-lifting system near the tank, opposite the embankment. Fourth, locally available materials were employed in the construction of this barrage, such as dressed stones, rubble and earth, but nothing can be clearly said about the cementing material as further investigation is needed into this matter. Lastly, the reservoir resulted in the rise of the water table of the area, as indicated by the existence of many wells, step wells and *kunds* in the vicinity.

The extant structure of the Indra Sarovar Dam encourages us to infer that the hydraulic engineers at that early period had developed expertise to erect an embankment best suited to the available site and could give tilt to the projection. The weight on the dam too appears to have been kept proportionate to ensure safety of the dam. Similarly, the size of the steps was kept small knowing the level of force of the water. This embankment appears to be one of the early structures of the region, dating back to the ninth century.

NOTES

1. See Ghulam Yazdani, *Bidar: Its History and Monuments*, London: Oxford University Press, 1947, pp. 203–5; Iqtidar Alam Khan and Ravindra Kumar, 'The Mansagar Dam of Amber', in *Technology in Ancient and Medieval India*, ed. Aniruddha Ray and S.K. Bagchi, Delhi: Sundeep, 1986, pp. 25–38; Rajshekhar Vyas, *Mewar Ki Kala Aur Sthapatya*, Jaipur: Rajasthan Prakashan, 1988, pp. 221–4; Rajiv Sharma, 'Raj Samand Dam: An Achievement of the 17th Century Civil Engineering', *Proceedings of Indian History Congress*, 1992–3 (hereafter *PIHC*), pp. 219–24; Vinod Kumar Singh, 'A Chandella Reservoir: Kalyan Sagar of Mahoba', *PIHC* 1992–3, pp. 261–4; Anupam Mishra's two books, *Aaj bhi Khare Hain Talab*, Delhi: Gandhi Shanti Pratishthan, 1994 and *Rajasthan Ki Rajat Boonde*, Delhi: Gandhi Shanti Pratishthan, 1995; Anil Agarwal and Sunita Narain, *Dying Wisdom: Rise, Fall and Potential of India's Traditional Water Harvesting Systems*, Delhi: Centre for Science and Environment, 1997, pp. 104–75; Dominic J. Davison-Jenkins, *The Irrigation and Water Supply Systems of Vijayanagra*, Delhi: Manohar, 1997.
2. The tradition of constructing of wells, *baoris*, tanks and dams for religious purposes was widely prevalent in Rajasthan. For such waterworks in Mewar: Ranchhod Bhatt, *Amarkavyam*, ed. S.K. Sharma and R.P. Bhatnagar, Udaipur: Sahitya Sansthan, Rajasthan Vidyapeeth, 1985, pp. 81–3. For similar works in reference to Mewar: Munhota Nainsi, *Marwar-ra-Pargana-ri-Vigat*, ed. N.S. Bhati, I, Jodhpur: Rajasthan Oriental Research Institute, 1968, Appendix 1a, pp. 562–4.
3. Sacred and religious purposes included waterworks erected in memory of a deceased person, Rana Mokal constructed a tank for the salvation of his brother Rawat Bagha, cf. Ranchhod Bhatt: *Rajprashasti*, op. cit., p. 40.
4. *Dying Wisdom*, op. cit., p. 281. The statues of gods and goddesses were necessarily attached to the water reservoirs in Mewar. This was promoted by the belief that gods would protect their edifices.
5. The temple of Eklingji houses Lord Shiva. Since the time of Rawal Bapa, the rulers of Mewar consider Eklingji as the real master of the state and that is why they call themselves His *diwan*.
6. This was a part of the temple. Tryna Lyans writes that the Eklingji temple complex is situated in a valley 22.4 km. north of Udaipur,

the capital of the former state of Mewar (in Rajasthan). The site, also known as Kailashpuri, is the locus of various legends about the origin of Mewar's royal family.

This temple stands roughly at the centre of walled compound. Just over the wall behind the memorials to the deceased monks is a lake called Indra or Bhojela Sarovar. Cf. 'The Changing Faces of Eklingji: A Dynastic Shrine and Its Artists' in *Artibus Asiae*, vol. LVIII, ¾, Washington, D.C.: Smithsonian Institution, pp. 254–5.

7. Shyamaldas mentions Indra Sarovar and writes that it is also known as Bhodela, see his *Vir Vinod*, I, p. 1593. K.D. Erskine, too, notices, 'a picturesque lake' in the vicinity of the temple, see *The Mewar Residency*, op. cit., p. 106.
8. This survey was conducted with the financial assistance from the Centre of Advanced Study in History, Aligarh Muslim University. Zameer Ahmad, Senior Technical Assistant accompanied me and took measurements.
9. The locality was fashioned after the Lord Shiva's favourite mountain, Kailash, i.e. the personified peak of the Himalayas. In the eighth century, the Guhilots established their rule in this region and constructed the temple of Eklingji.
10. Its geographical location is described in the Kumbalgarh inscription of VS 1517/AD 1460. A.K. Vyas, 'First and Third Slabs of Kumbalgarh Inscription of VS 1517' (hereafter Kumbhalgarh Inscription), *Epigraphia Indica*, vol. XXIV, 1937–8, pp. 304–6; Nainsi, *Khyat*, I, p. 34; Ojha, *Udaipur Rajya*, op. cit., p. 336; K.D. Erskine, op. cit., p. 106.
11. Erskine, ibid., p. 5.
12. James Tod, *Annals*, op. cit., p. 410 describes its location thus: 'The hills towering around it on all sides are primitive formation and their scrap summit are scattered with honeycombs'.
13. Ibid., p. 410.
14. For the course of the Banas, Shyamaldas, op. cit., I, p. 110; Erskine, op. cit., p. 7.
15. The Nathdwara stands on the north-east of Udaipur and is the principal Vaishnava temple: H.B.W. Garrick and A. Cunningham: *Report of a Tour in the Punjab and Rajputana*, vol. XXIII, Delhi: Indological Book House, rpt., n.d., pp. 99–101.
16. Kumbhalgarh Inscription, vss. 18–19, p. 306.

17. The dam is located at the back of the temple and the distance between the two is about 80 m.
18. This conjecture receives support from the statement in *Vir Vinod*, I, p. 159, that both the establishments are contemporary.
19. An interesting mythological story is widely current in Mewar and recorded in different sources. The tale records that this tank is said to have been in existence since the time of Lord Indra, the king of Hindu *devatas*. He mediated at the feet of Eklingji and dug out the valley to carry out a penance imposed on him for killing a Brahman. Thus, the tank came into existence and was called Indra Sarovar, Kumbalgarh Inscription, vs. 25–8, p. 306.
20. Vyas, op. cit., 1984, p. 305. G.N. Sharma, *Social Life in Medieval Rajasthan*, Agra: L.N. Agarwal, 1968, p. 187.
21. *Amarkavyam*, p. 83.
22. Vyas simply writes that he was one of the early predecessors of the Guhil family, op. cit., p. 305, while the *Amarkavyam*, pp. 84–5 gives VS 166/AD 109 [?] as the last year of the reign of Kal Bhojaditya. It does not seem to be correct; it should have been VS 766/AD 709, cf., *Vir Vinod*, I, pp. 219–54.
23. Ojha, op. cit., pp. 404–10.
24. *Vir Vinod*, I, pp. 250–2.
25. On another slab, the name is recorded in this fashion: '*Sutradhar Kanoji Josi Jadan—Sutradhar Chantra*'.
26. In original it's a long inscription but I could decipher it only partially; it goes as follows: '*Pal bandhya Tara sammat 1202* [?] *pagoti dijiyo. . .*'
27. AD 1380.
28. *Rajprashasti Mahakavyam*, pp. 110–11. The writer of the epic says that he had composed an eulogy (*prashasti*) on the Indra Sarovar and was ordered by the Maharana to engrave it on a stone slab. Unfortunately, this inscription is not traceable. According to Shyamaldas, the repair was carried under Maharana Mokal (VS 1454/AD 1397—VS 1490/AD 1433) and Maharana Raj Singh, *Vir Vinod*, I, p. 159. While Rajshekar Vyas draws a conclusion from the same verses of the *Rajprashasti Mahakavyam* to the effect that instead of the old structure a new one was erected by Maharana Raj Singh. This does not seem to be correct (cf. Vyas, op. cit., p. 233).

29. *Rajprashasti*, pp. 110–11.
30. There is an inscription on the outlet which shows that it was probably constructed in VS 1767/AD 1710. It means that originally it was not planned at the time of construction. Though it does not appear to be a sound conclusion but we have to accept it until new evidence comes to light.

    The names of the inscribers are Sutradhar Kisno Purabi and Kisno Sompura. Lyons has collected interesting information about the architect caste of Sompura, see op. cit.
31. This garden is an enclosure, measuring 50 × 50 m.
32. *Rajprashasti*, pp. 110–11.
33. Donald Hill, *A History of Engineering in Classical and Medieval Times*, London: Croom Helm, 1984, pp. 48–9.

CHAPTER 3

Udai Sagar Dam

The construction of waterworks appears to have become a part of the cultural ethos for the rulers of Mewar (Rajasthan). Innumerable extant structures are witness to it.[1] It is also reflected in literature and inscriptions composed in the early and late medieval times.[2] In fact, these works are the true memorials of the builders. Following the family tradition, Maharana Udai Singh (AD 1540–72) planned a big project encompassing the founding of a new capital city and the erection of a dyke (*pal*) for irrigating the surrounding agricultural tract.[3] The site selected for it was located to the east of the city. The existence of a deep gorge between the hills was fancied by the Maharana as an ideal location for the construction of a weir. He successfully achieved his aim by creating a large lake (*sagar*) by joining the two isolated hills with a wall of rubble with line mortar which has attracted both contemporary and modern writers.[4]

For our study, we made an extensive survey of the sixteenth-century Udai Sagar Dam and took detailed measurements of its every part. On the basis of this survey we have been able to draw its ground plan[5] (Figure 3.1). The aim, however, is to study the structure in its entirety. An attempt has been made to determine its date of construction, explore the canals by which the vast agricultural tract was irrigated, and which furnished substantial income to the state.

I

The Udai Sagar Dam is located about 12 km. east of Udaipur near the valley of Debari.[6] There was a cluster of hills (*dungar*) here which was surrounded by villages owned by the Devra

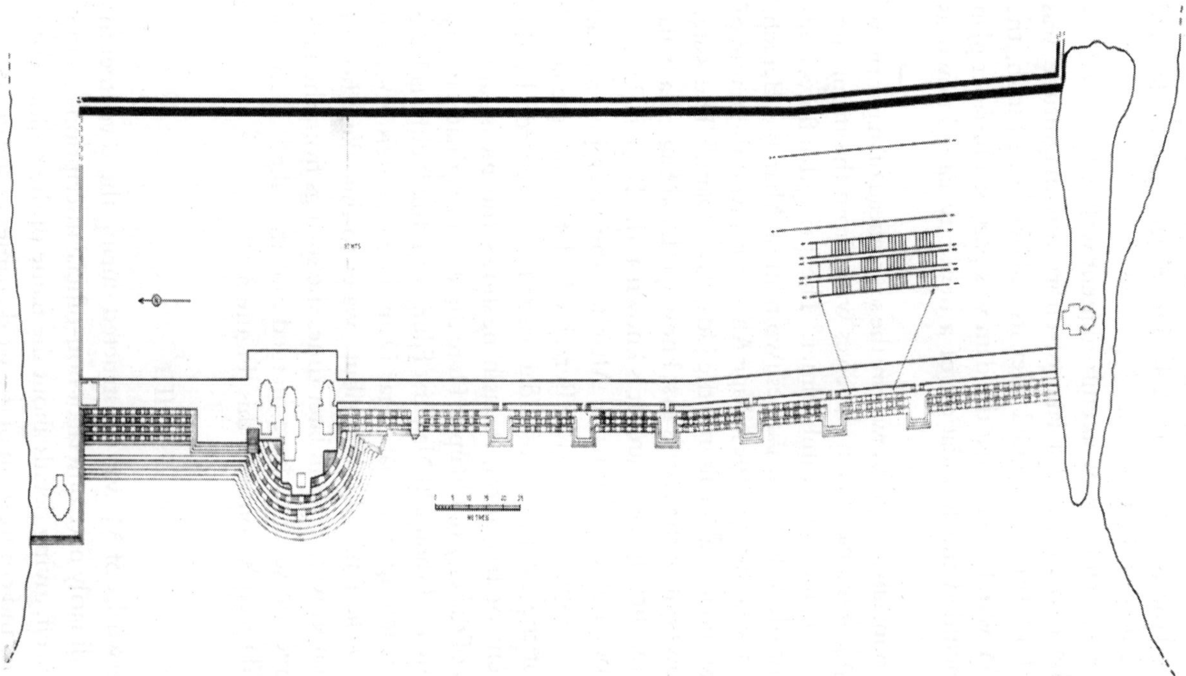

Figure 3.1: Ground Plan of Udai Sagar Dam

Rajputs, the *bhumias* of the area.[7] The maximum height of the hills around the area reaches to 234.85 m. A deep gorge is formed by the mountains (Figure 3.2).

Here at a suitable point an embankment wall was erected between the two hills and a lake of enormous magnitude was created. The dam was named after its builder Udai Singh, the ruler of Mewar. It is said that the dam was part of a broader plan which included the foundation of a new city later known as Udaipur.[8]

The mountains around Mewar are the sourcehead of major rivers. The deep gorge near Debari receives water from the mountains (*magra*) of Gogunda and Kumbhalmer.[9] The spring that flows from the hills north of Udaipur carries two names: Ahar and Berach. When it passes through the village Ahar, it is given the name of the village and it finally falls into the Udai Sagar Dam.[10] The water which flows out of the lake is first known as Udai Sagar ka Nala. When it reaches the open area it is known as the Berach River.[11] Nainsi does not use the name Ahar for the river but instead designates the spring as the river Berach which falls into the lake.[12] A chart of the catchment of Udai Sagar Dam prepared by the Department of Irrigation, Rajasthan, shows the mountains to the north of the city of Udaipur. Here the water originates from a place known as Gorana ki Nal, first falling into the Pichhola Lake and then reaching the Udai Sagar Dam. The channels from the *Bari* and *Madar* tanks also contribute water to the lake. Though both the tanks were built in later times the springs from the hills in this direction seem to have merged into the lake.[13] A survey conducted recently confirms this (Figure 3.3).

III

The waterworks of Mewar constructed during the seventeenth century ordinarily have official or non-official inscriptions on the structure itself, giving details about the name of the builder, date of initiation and completion of the work, names of the engineers (*sutradhara*) and masons and sometimes the cost incurred on the

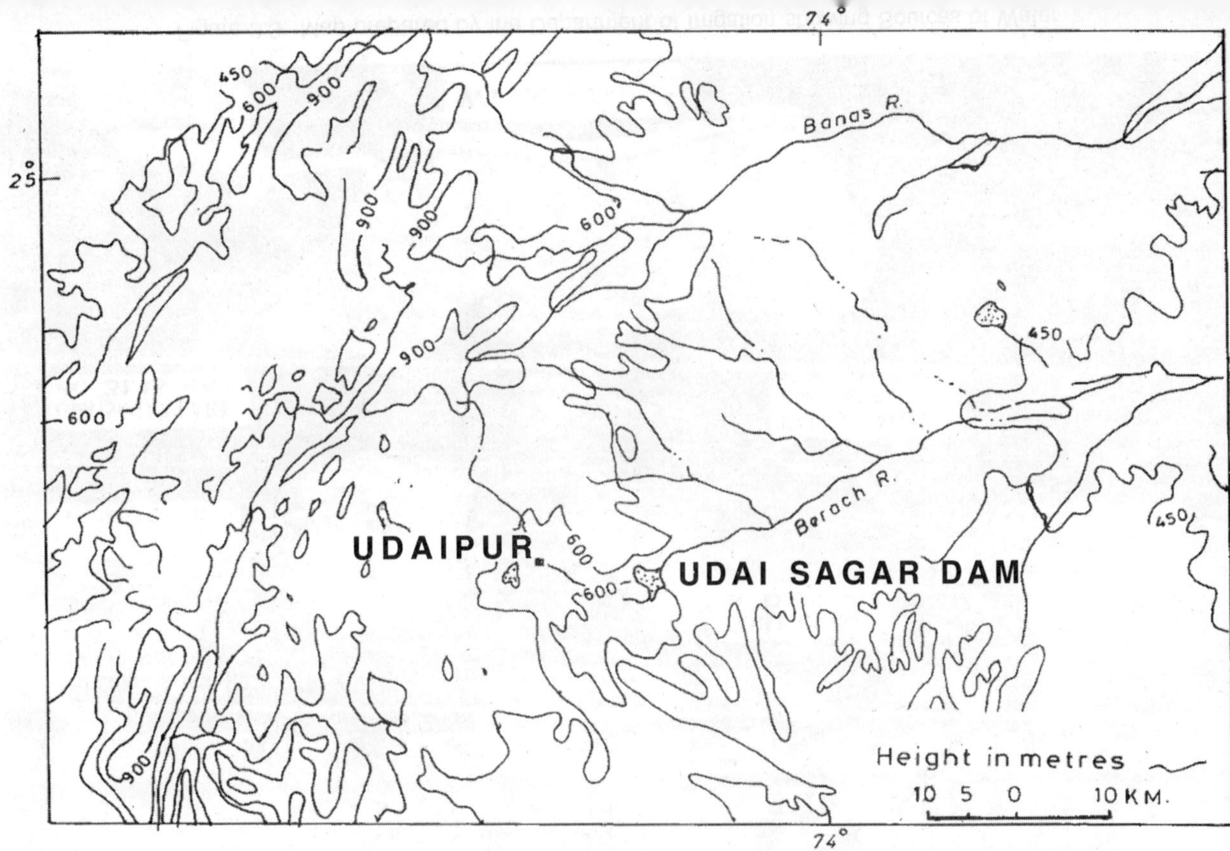

Figure 3.2: Udai Sagar Dam and its Environs

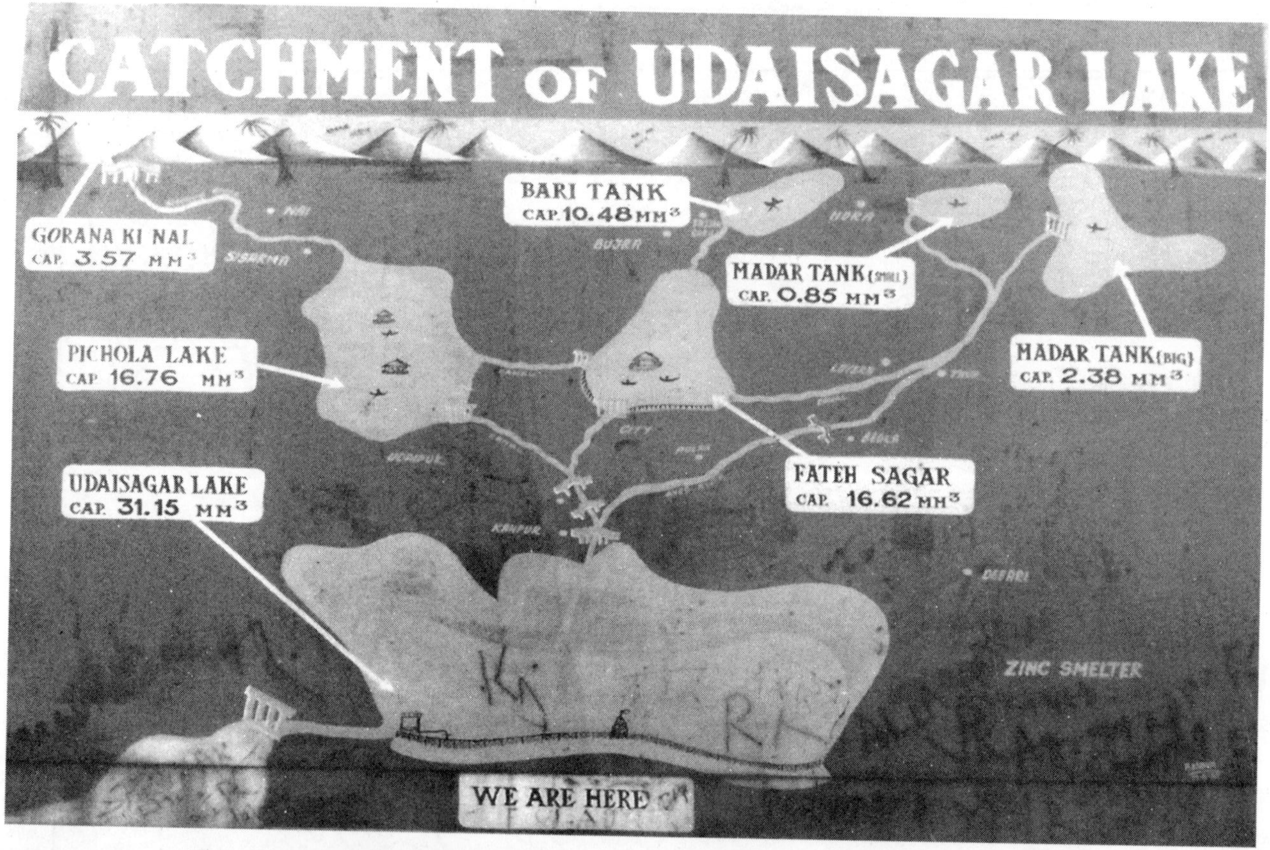

Figure 3.3: Map prepared by the Department of Irrigation showing Sources of Water

structure.[14] But unfortunately the Udai Sagar Dam does not have an official inscription by its builder. This has led to speculation about the date and time of its completion. There is no unanimity among the historians about the date of construction of the Udai Sagar Dam. The uncertainty is reflected in Nainsi's *Khyat* in which he gives two dates VS 1620/AD 1563 and VS 1621/AD 1564—for the building of the structure.[15] Ranchhod Bhatt composed two epics—*Rajprashasti Mahakavyam* (VS 1718/AD 1661) and *Amarkavyam* (VS 1748/AD 1683–VS 1750/AD 1693). In the former, he does not mention the date of construction while in the latter, the year VS 1600/AD 1543 is given as the beginning of the construction and VS 1618/AD 1561 as the completion of the dam.[16] Another seventeenth-century source mentions the year VS 1618/AD 1561 as the year of completion of the dam.[17]

The celebrated historian of Mewar, Shyamal Das, records the year VS 1616/AD 1559 as the inauguration of the work and VS 1619/AD 1562 as the year of completion.[18] Ojha agrees with Shyamal Das regarding the date of the beginning of the work on this dam and gives the year VS 1621/AD 1564 for its completion.[19] James Tod merely writes that the lake was formed several years previous to the catastrophe, i.e. before 1568.[20] This kind of uncertainty prevails in other sources too.[21] However, this uncertainty is removed by the recently discovered contemporary source, the *Chittor Udaipur Patnama* which records the year VS 1615/AD 1558 as the initiation of the work,[22] which means that the dam was constructed well before the capture of Chittor Fort by Emperor Akbar.

This fact gets further support from the date discovered on the structure. We noticed the stone-cutters' marks and signatures of the masons, sometimes with date, on the temples erected at the top of the dam. On one of the temples, the year VS 1623/AD 1566 is inscribed by one of the masons along with his signature.[23] This date assumes significance. The temples were built after the completion of the water structure. It shows then that the dam was ready at least 6–7 years prior to this date, that is in 1559 or 1560. This date confirms the time of the completion of the structure.

IV

The Udai Sagar Dam is a reservoir of enormous capacity. High hills along the four sides create a deep depression. The river enters into this valley from the south-eastern side near the village Lakadwas. The passage of water is blocked by erecting a masonry wall between the two hills which run north to south. On the opposite side facing the wall is a continuous range of hills which makes this gorge like a deep basin (Figure 3.4).

Abul Fazl was the first chronicler who noticed the existence of a large lake at Udaipur with a circumference of about 64 km. Wheat crops were being grown by the waters of the tank.[24] Celebrated historian and statistician of seventeenth-century Rajasthan, Nainsi, too, gives the circumference of the tank as 40 km.[25] At another place he writes that when the water from the hills of the Gogunda and Kumbhalmer join to form the river Berach and water gushes into the tank, its circumference expands and reaches up to 80 km.[26]

The information provided by Abul Fazl and Nainsi are similar. From this, it becomes clear that the structure identified in the *Ain* is certainly that of the Udai Sagar and not the Jai Samand as erroneously recognized by an eminent historian.[27] The dam noticed by the Mughal chronicler is located at Udaipur, while the Jai Samand is in the *pargana* of Tana.[28] It is located towards the south-east and is 51.2 km. from the city of Udaipur.[29]

Nainsi further records the measurements of different sections of the dam. He gives the length and breadth of the *pal* as 305 and 152.5 m. respectively.[30] The total height recorded comes to 42.7 m.—35.4 m. of the wall was in water while 7.3 m. was exposed. Then, he enters the size of the canal (*nala*). Its depth is recorded as 30.5 m. and the breadth as 7.3 m. and it was cut out of the hill (*bhakhar*).[31]

The measurements provided by Nainsi require verification. No one has compiled the data in so much detail. After a gap of 350 years or so, Erskine records only the breadth of the structure as 55 m.[32] Ojha repeats him while Raj Shekhar Vyas gives the

Figure 3.4: Front View of Embankment of Udai Sagar Dam

length and the breadth as 30.5 and 55 m. respectively.[33] But none matches with Nainsi's observations.

We took up extensive physical survey of the structure. Detailed measurements of each of the section was carried out. On this basis, a ground plan was prepared which is given in the beginning. Detailed photographs from different directions were also taken.

The wall erected between the two hills runs north-south. The obvious aim was to hold back water or to conserve water to channellize it in the desired direction. The length measured comes to 303 m. The engineers took extra care about the safety and long life of the dyke. This necessitated building of the back or retaining wall to muster additional strength to the embankment wall. The distance between the two walls comes to 97 m., which is the width of the embankment. A width of 1.20 m. was given to the retaining wall. The space or gap between the front and back wall was filled with lime and mortar.

Protection and safety appear to have been of utmost concern to the civil engineers. The danger of washing away of the structure in the rainy season was always looming. In order to obviate this eventuality, the embankment wall was imparted strength by additional projections. The maximum breadth in the north comes to 28 m. and decreases as we move towards the south. Possibly the northern side wall faced the violent thrust of water gushing from the south-eastern direction. Interestingly, the projections were in descending order towards the water.

In the north is a pathway adjoining the hill that goes towards the other side of the dam. The embankment wall starts from here. On this a rectangular bastion has been erected and a circular temple built. Adjoining the bastion, the wall of the first section extends towards the south and joins the semi-circular projection towards the lake (Figure 3.5). The section has seven projections in descending order. Stairs are provided on both the sides to reach the fourth floor. A second section starts in the form of a semi-circular projection. Its width from the embankment wall to the water comes to 28 m. It has ten platforms. The access to the sixth platform is provided by steps.

Figure 3.5: Semi-Circular Projection of Udai Sagar Dam

The embankment wall from the semi-circular bastion towards the south is further divided into six sections by constructing solid platforms of 4.80 × 4.00 m. The water-lifting place served as the seventh platform. This is located on the north near the semi-circular projection. Its size is 6.50 × 2.40 m. Five stairs 1 m in length join the two platforms. The division of the projections into sections was certainly meant to impart maximum strength to the embankment wall (Figure 3.6).

Towards the southern corner of the dam, a temple similar to the one in the north has been built. In the same direction stands a guard's chamber near the outlet (Figure 3.7). Two objectives seem to have been in operation behind erecting this structure at this point: first, to control the supply of water in the canal, and, second, to look after the safety and protection of the dyke and take necessary steps at the time of danger, particularly during torrential rains.

An important component part of the water structure is the outlet located at the end of the embankment in the south. Nainsi rightly observes that the channel (*nala*) was cut out from the mountain. He records its size as 30.5 m. deep and 7.3 m. in breadth.[34] The round figures perhaps indicate that these were based on conjectures. Our measurement of the outlet came to 4 m. in width (Figure 3.8).

At the end of the outlet, a rectangular cistern has been constructed to store water released in the canal (Figure 3.9). At the both ends of the outlet are devices to control the supply of water into the cistern. Two canals convey water into the cistern and two more supply water towards the south-west and the north. The former goes first towards the west and then turns southward (Figure 3.10), while another flows towards the north and goes parallel adjoining the back wall and then takes a turn towards the west (Figure 3.11). Two floodgates regulate water supply to the canals. Both the channels are of masonry and irrigated a large area where, as the *Ain* notices, wheat cultivation was done. Its width is 1.15 m. Its length could not be measured, but it appears to be quite long in size. It is noteworthy that at regular intervals

Figure 3.6: A View of Sub-section of the Wall

Figure 3.7: A View of Guard's Chamber

Figure 3.8: Outlet in the South

Figure 3.9: View of Rectangular Cistern

Figure 3.10: Canal on Back of Udai Sagar Dam Bordering Hill

Figure 3.11: View of North Canal Parallel to the Back Wall

square or rectangular outlets of 0.61 × 0.61 m. were cut for the drains going to the agricultural fields (Figure 3.12).

Near the back wall, a sloping passage was meant to go to the rear of the dam. Along the pathway a masonry wall was constructed. The passage at the end of the slope trifurcates. The right and left directions lead towards the canal. The straight path goes towards a residential complex meant probably for the officer-in-charge of the dam.

Niches are carved out on half of the embankment wall towards the southern direction. Stone slabs are on the four sides of these niches. One of the slabs is signed by a certain artisan named Parsa. The reason for carving out the niches is not clear. Was there any technical reason behind it, or was it simply a design pattern?

V

The study of the Udai Sagar Dam demonstrates that the site chosen by its builders and civil engineers was perfect. It was surrounded by hills with only one opening which was closed by a masonry structure huge in length and breadth. Both dimensions increased its weight which was considered the strength of the dam. This was the exclusive technique employed against the pressure of the water. It is the weight of the structure which has enabled it to remain standing for 450 years coping with the vagaries of weather. No damage has been done to the dam to this date.

The hydrological factor was the dominating element in the dam's plan. The purpose behind the construction of this dam was irrigation. The builder wanted to create a vast fertile tract to feed the residents of the newly-founded city. He got tremendous success in his mission by increasing wheat cultivation. This embankment is still active, irrigating the surrounding fields. The report of the Department of Irrigation, Rajasthan, shows that 6,070 acres of land was being irrigated during 1992–3. One can conclude on this basis that the Udai Sagar Dam is an excellent example of 'gravity' dam constructed during the middle of the sixteenth century. The refinement and improvement in the indigenous dam technology can be inferred from this study.

Figure 3.12: View of Square Rectangular Outlet in Canal

NOTES

1. In one source, it is mentioned that the Udai Sagar is the eighth lake which denotes the existence of seven more lakes previous to its construction by Maharana Udai Singh (see 'Sisod Vansavali Evam Rajasthan ke Rajwadon ki Vansavaliyan', in *Majjhamika*, vol. 23, Udaipur: Pratap Shodh Sansthan, 1995, p. 55). Raj Samand, Pichhola and Fateh Sagar are the well-known names (cf., *Rajprashasti;* Nainsi, *Khyat,* I, pp. 32–3; *Chittor-Udaipur ka Patnama*, op. cit., pp. 244–9; Erskine, op. cit., vol. VIIA, pp. 6–10).
2. The first stanza of the *Rajprashasti Mahakavyam* makes it clear that the rulers of Mewar took inspiration for building of bridges or dykes from Lord Rama. Bridge erection brought popularity for him in this world (p. 1). Besides this, the Sisodia rulers like all Hindus, had deep faith in the theory of salvation (*moksha*). The construction of tank brought religious merit to the ancestors of the builder as well as to himself. Gradually, this became a part of the cultural ethos of the rulers. That is why almost all the Maharanas endeavoured to erect water structures in their lifetime (cf. *Rajprashasti Mahakavyam*, pp. 40–1. References are scattered all over the text).
3. *Chittor Udaipur ka Patnama*, pp. 244–9.
4. *Ain-i Akbari*, II, p. 278; Nainsi, *Khyat*, I, pp. 32–3; James Tod writes that 'The dam thrown across a gorge of mountains is of enormous magnitude and strength, as is necessary, indeed, to shut in a volume of water twelve miles in circumference'. *Annals and Antiquities*, pp. 263 and 619.
5. Zameer Ahmad, Technical Assistant (draftsman) and Ghulam Mujtaba, photographer of the archaeological section, accompanied me in the survey work and prepared the plans.
6. Nainsi, *Khyat*, I, p. 35; 4 km. is approximately equal to 2½ miles Erskine too gives the location of the Udai Sagar as 12.8 km. east of Udaipur (op. cit., p. 9).
7. *Chittor Udaipur ka Patnama*, pp. 242–7.
8. Ibid.
9. Nainsi, *Khyat*, I, p. 35.
10. Erskine, op. cit., pp. 7–8.
11. Ibid. The river Berach is a tributary of the river Banas, the biggest river of Mewar.

12. Nainsi, *Khyat*, I, p. 35.
13. See the photograph.
14. Raj Samand has perhaps the longest inscription in India running into 25 block stone slabs of 0.9 × 0.76 m. giving each and every detail about the dam. The first slab contains the prayers to gods and goddesses. The remaining 24 slabs have one section of the 'epic' poem engraved on each of them. It was composed by Ranchhod Bhatt and published under the title *Rajprashasti Mahakavyam*. Similarly, there is an inscription on the parapet of Jana Sagar or *Bari ka Talao* (see Appendix 2 in *Rajprashasti*, pp. 272–7). The Jai Samand or Dhebar also carries inscriptions.
15. Nainsi, *Khyat*, I, p. 34.
16. *Rajprashasti*, p. 41; *Amarkavyam*, op. cit., pp. 213–14.
17. 'Sisod Vansavali Evam Rajasthan ke Rajwadon ki Vansavaliyan', op. cit., p. 55.
18. *Vir Vinod*, II, part I, p. 73.
19. *Udaipur Rajya ka Itihas*, p. 311.
20. Tod, op. cit., I, p. 263.
21. Erskine says that the embankment was built between 1559 and 1565 (see, op. cit., p. 9. Raj Shekhar Vyas follows Erskine (see *Mewar ki Kala aur Sthapatya*, p. 128).
22. *Chittor Udaipur ka Patnama*, pp. 241–8.
23. The date is engraved on the stairs of the temple in the north. This slab is signed by one Shri Balaji. On another site, the words inscribed are *Shri Likhtam Patel Kachra Jamburam* ('Patel Kachra writes'. The word 'Shri' is generally employed in the beginning of any work). Similarly, two masons, Bhat Mohan and Vachali Ramo Viko also inscribed their names.
24. *Ain-i Akbari*, II, p. 278.
25. Nainsi, *Khyat*, I, p. 34.
26. Ibid., p. 35.
27. Irfan Habib in the first edition of his book (*The Agrarian System of Mughal India*, Bombay: Asia Publishing House, 1963, p. 28) rightly identified the structure in the *Ain* as the Udai Sagar. But in his *An Atlas of the Mughal Empire*, Delhi: Oxford University Press, 1982, p. 19, he changed his earlier view and identified it as the Jai Samand, basing his argument on the size of the lake. He writes that the size of the dam shown in the *Ain* is only that of the Dhebar Lake. Again, he asserts this view in the second (revised) edition of his book

(*Agrarian System*, Delhi: Oxford University Press, 1999, pp. 30–1). His entire argument is based on James Tod's description (cf. *Annals and Antiquities*, I, p. 619). But he has failed to notice two evidences. Nainsi's two statements about the circumference of the Udai Sagar, and the location of the Jai Samand which is 51 km. away from Udaipur.

28. *Maharana Raj Singh-Pargana Bahi*, op. cit., p. 10.
29. Ojha, op. cit., pp. 309–10.
30. Nainsi, *Khyat*, I, p. 34.
31. Ibid.
32. Erskine, op. cit., pp. 8–9.
33. Ojha records the measurements of the structure. According to him it is 4 km. in length and 3.2 km. in breadth. Its catchment area is 473.6 sq. km. The dam is fed by river Ahar. The embankment wall is constructed between the two hills and is quite high; its breadth is 55 m. (cf. *Udaipur Rajya ka Itihas*, I, p. 7); Raj Shekhar Vyas, *Mewar ki Kala Aur Sthapatya*, p. 128.
34. Nainsi, *Khyat*, I, p. 34.

CHAPTER 4

Raj Samand Dam

Maharana Raj Singh will be remembered in the history of hydraulic works for constructing the Raj Samand Dam in 1671. It is at a distance of about 60 km. north of Udaipur. It is not only massive but also beautiful. The northern portion of the dam is entirely faced with white marble and the pavilion is adorned with carved sculptures. This portion is called the *Nauchauki* (nine platforms). On both ends stand palaces. On the northern side, the palace adjoins the dyke while the buildings in the southern direction are located at the top of a hill. An enclosed chamber stands on the embankment which was probably meant for the bathing of royal ladies. The northern side is enclosed by walls with an imposing gate in the middle. This shows that entry was prohibited to public but open for members of the royal family and the staff working on the structure. This portion formed a beautiful retreat for the Maharana and his entourage. Other portions of the structure were open to public which is attested by large settlements of people around them. Thus, two objectives were behind the construction of this dam: irrigation and a pleasure retreat for the royal family.

Raj Samand is perhaps the first dam which tells its own story in great detail, from its origin to its completion. Sanskrit verses were composed by a court-poet on the command of its builder Maharana Raj Singh. Later on, a long composition was inscribed on the slabs and fixed in the niches on the upper part of the structure made specifically for the purpose.[1] This portion of the dyke is called *Nauchauki*.

The *Rajprashasti* is not only a history of the Sisodia ruler of

Mewar but also a dam-centric treatise. At least 12 chapters (from the ninth to the twentieth *sarg*) contain information on this water harvesting structure. The entire process of its construction from the foundation till its completion is recorded here. It offers technical details about the digging of the foundation and the method of draining out water from the site. It records the measurement of every part and section of the structure along with the material used in it. The management and organization of the work also find place in it. The division of work among the officials with full financial responsibilities is carefully entered. We also get information on the engineering aspects. It records the break-up of the expenditure incurred on the dam. In fact, it is a unique treatise on the history of dam technology in medieval India, specifically in Rajasthan.

I thrice undertook a physical survey of the dyke along with a team from the archaeology section of our department.[2] The data regarding the dimensions of each part of the barrage obtained enabled us to make a comparison with those available in the *Rajprashasti*.

ORIGIN OF THE DYKE PLAN

Two different reasons are assigned by the contemporary authorities about the origin of the idea for the construction of the dyke. Ranchhod Bhatt, author of the *Rajprashasti*, writes that the seed of this plan was in the mind of the Sisodia ruler since the time he was a prince. When in VS 1698/AD 1641 he was passing through village Kankroli en route to Jaisalmer for his wedding, he happened to see a site suitable for the lake.[3] After 20 years, when he ascended the throne in VS 1718/AD 1661, Raj Singh again visited the site and made up his mind to build a dyke there.[4] The Maharana sought the advice of the family priest on the construction of the dam (*setu*).[5] The priest supported the plan but on three conditions: first, the construction should be according to the will of the person; second, it should be carried out without dispute with the emperor of Delhi; and third, the construction requires enormous wealth which should be ensured first. The Sisodia ruler accepted his priest's advice and told him that all the preconditions were within his reach.

Another contemporary authority, a court-poet, Maan assigns the devastating famine of VS 1717/AD 1660 as the main reason behind the construction of the Raj Samand. He writes that the scarcity of grain created havoc, compelling parents to sell their children and husbands to abandon their wives. To protect his people from the vagaries of famine the Sisodia ruler took the decision to build a large dam.[6] Thus, this waterbody came into existence as a famine relief work which provided employment to thousands of people during the period of distress.

Various other reasons have been offered by contemporary sources which were compiled in the *Vir Vinod*. It is said that once the movement of the Maharana was blocked by the heavy flow of water in the river Gomti due to torrential rains. This led the Maharana to think of building a barrage on the river to control its fury.[7] A second, reason given is that it was undertaken as a penance for the murders of a Barhat, a *purohit* (priest), and a queen by the Maharana.[8]

GEOGRAPHY AND ITS LOCATION

The Raj Samand is located about 60 km. north of Udaipur in the vicinity of Kankroli. A cluster of hills ranging from 560 to 600 m. high in the north-west and north-east directions had created a deep depression in a wide open field (Figure 4.1). The river Gomti flows from the northern hills and crosses the hills located in the south and west. At this point its movement is blocked by erecting an embankment wall. It is the chief dam located in the north; other segments are projected towards water. In real terms it is not a straight barrage between two hills. On the other side of the hill it further extends in a semi-circular form. The northern portion of the dam is decorated with white marble. The total length of the dam is about 4.8 km.

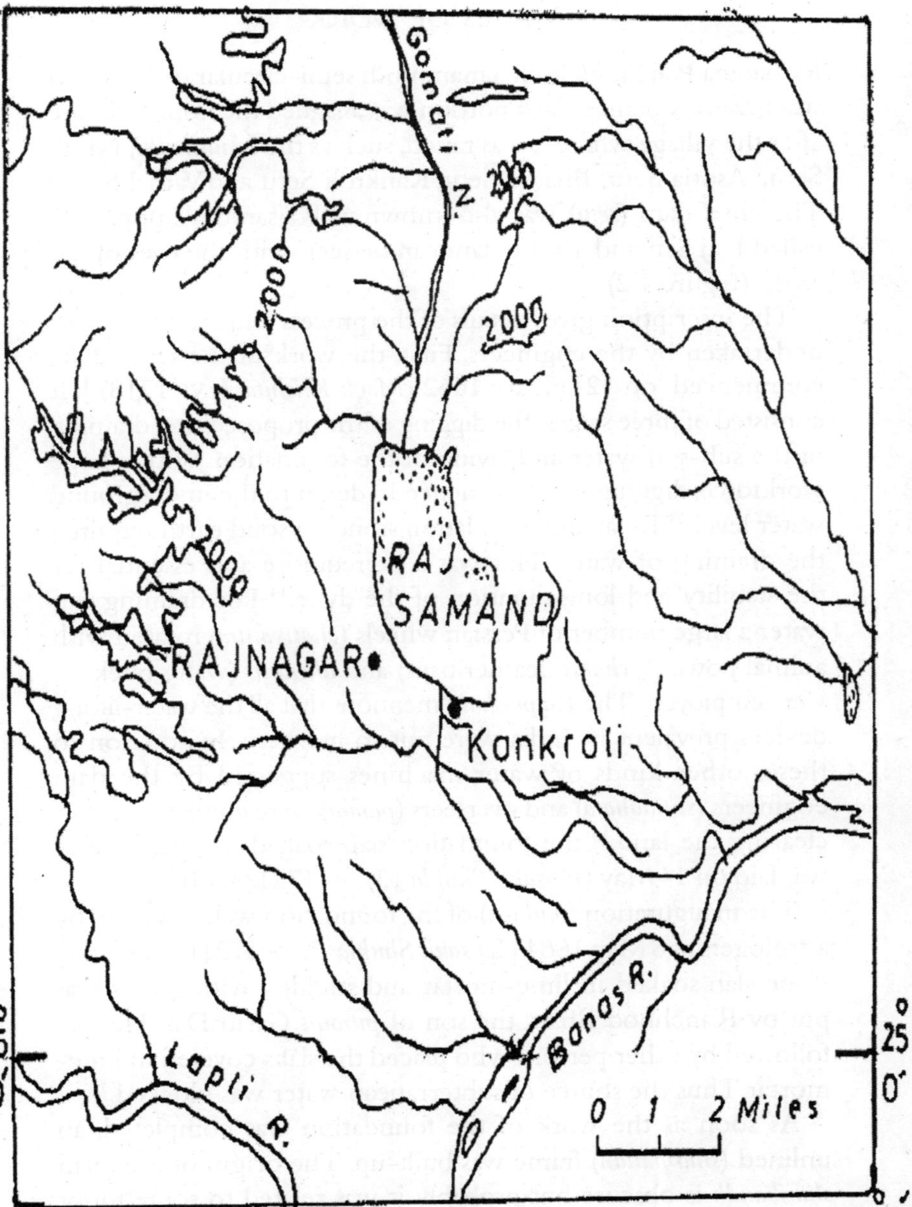

Figure 4.1: Raj Samand Dam and its Environs

STAGES OF WORK

Maharana Raj Singh built a mammoth semi-circular dyke which was given six names. Each portion was assigned the nomenclature after the village where it was raised, such as the *Nauchauki*, Nimb Setu, Asotia Setu, Bhadra Setu, Kankroli Setu and Vansal Setu.[9] The chief dam (*setu*) was also known as Rajsamdra[10] popularly called Raj Samand. Entire dam can be seen with the eyes of the artist (Figure 4.2).

The inscription gives details of the process and stages of work undertaken by the engineers. First, the work of the chief dyke commenced on 12 January 1662 (*Magh Krishna* 7, VS 1718).[11] It consisted of three stages: the digging of the proposed site, draining of the sub-soil water and laying of the foundation. With a large workforce, digging was commenced[12] down to the underground water level.[13] To lay the foundation stone on solid earth required the draining of water. This was a prerequisite and essential for the stability and long duration of the dyke.[14] For draining out water a large number of Persian wheels (*ghatiyantra*) rotated with animal power,[15] *charas* (leather bags) and *dhenklis* (water buckets) were employed. The *Rajprashasti* mentions that all the water-lifting devices prevalent in India were put to use here. In addition to these, other kinds of water-machines suggested by the dam engineers (*sutradhara*) and overseers (*pamar*) were employed. After clearing the land,[16] the foundation slab soaked in lime-mortar was laid on 14 May (*Baisakh Shukla* 13), VS 1721/AD 1664.[17]

The inauguration (*muhurt*) of the foundation wall was fixed by astrologers as 8 May 1664 (*Baisakh Shukla* 7, VS 1721).[18] The first stone slab soaked in lime-mortar and studded with jewels was put by Ranchhod Bhatt, the son of *purohit* Garib Das. He was followed by other persons who placed the slabs covered in lime-mortar. Thus the source of subterranean water was blocked.[19]

As soon as the work of the foundation was completed, an unlined (*sharibandh*) frame was built-up. The origin of the term *sharibandh* is obscure but probably it was related to some forest material such as wood, leaves and rope. It would be safe to infer

Figure 4.2: Artist's View of the Entire Dam of Raj Samand

that the engineers first made a wooden structure using grass, leaves and rope and then the masonry structure was raised over it.[20] The *Rajprashasti* praises the master-builders who erected the structure with stone and lime-mortar, providing it strength.[21] Bhatt further writes that the builders had excavated a tank by erecting a dyke (*setu*).

The embankment wall was erected on the foundation which took eight years, five months and six days to build. The rain in the month of June (*Jyestha*) brought water into the tank which was drained out by the diggers. Then the middle part of the dyke was filled with the earth dug out from the square platform (*chatuspadi*). The *Rajprashasti* records people's participation in digging the *chatuski* in the tank.[22]

Rain water was collected in the tank and a boat was floated. On the *Jyestha Shukla* 15, VS 1728/AD 1671 the channel (*nala*) was blocked by the *sutradhars* to retain water.[23]

Cenotaphs of white and glazed marbles were constructed. The marble slabs were joined together with lime-mortar. The artisans and sculptors continuously worked and decorated it with carved patterns.[24] This part of white marble is called *Nauchauki* (Figure 4.3).

Construction on the second portion of the dam was initiated in village Kankroli. Here the *muhurt* for the foundation was on *Baisakh Shukla* 13, VS 1726/AD 1669[25] (Figure 4.4). The ground was dug out with numerous large tools and then filled with stone slabs dressed in lime-mortar. The dates of other four segments of the structure are not mentioned, perhaps they were part of the Kankroli dyke, or left out. Different nomenclatures appear to have been assigned to them because of their location in different villages.

Chapter (*sarg*) 11 offers details about the measurements of each part of the water body. The dimensions of all the six structures are recorded in *gaz* (Pers. *gaz*, a unit of measurement).[26] The author gives the length and breadth of the embankment wall, both of the foundation and also the part above the surface. Similarly, the height is also recorded with its break-up. The length and breadth of the back wall also are mentioned in the treatise. The size and

Figure 4.3: View of Main White Marble Dam of Raj Samand

Figure 4.4: View of Kankroli Embankment

extent of all the remaining portions of the structure are given. Faithful measurements of each section of the structure are remarkable and deserve appreciation. The measurements recorded along with conversion into metres are given in following Table:

Different Parts	Measurement	
	in *gaz*	in mts.
DIMENSIONS OF THE CHIEF DAM		
Length of foundation	515.00	314.15
Length above the surface	581.00	354.41
Breadth of foundation	55.00	33.55
Breadth above surface	10.00	6.10
Total height	57.00	34.77
1. Height of foundation	22.00	13.42
2. Height above surface	35.00	21.35
(a) Platform above the land	8.00	4.88
(b) *Mekhlas*	1.50	0.92
(c) *Tilak*	12.50	7.63
(d) *Sthawar*	13.00	7.93
Length of the back wall	700.00	427.00
Breadth of the foundation	18.00	10.98
Breadth above surface	5.00	3.05
Height of the back wall	28.00	17.08
DIMENSIONS OF THE NIMB SETU		
Length	432.00	263.52
Breadth of foundation	15.00	9.15
Breadth above surface	5.00	3.05
Height	10.00	6.10
DIMENSIONS OF THE BHADRA SETU		
Length	144.00	87.84
Breadth of foundation	12.00	7.32
Breadth above surface	5.00	3.05
Height	13.00	7.93

DIMENSIONS OF THE KANKROLI SETU		
Length of foundation	550.00	335.50
Length above surface	756.00	461.16
Breadth of foundation	35.00	21.35
Breadth above surface	7.00	4.27
Height of foundation	17.00	10.37
Height above land	38.00	23.18
Length of back wall	1,000.00	610
Breadh of foundation of the back wall	15.00	9.15
Breadth above surface of the back wall	10.00	6.10
Height of back wall	38.00	23.18

DIMENSIONS OF THE *SETU* NEAR VILLAGE ASOTIA		
Length	2,068.00	1,261.48
Breadth of foundation	18.00	10.98
Breadth above land	7.00	4.27
Height	24.00	14.64
Length of back *setu* in foundation	1,300.00	793.00
Breadth on top	10.00	6.10
Height	5.00	3.05

DIMENSIONS OF THE *SETU* NEAR BANSOL		
Length	1,224.00	746.64
Breadth of foundation	18.00	10.98
Breadth at the top	5.00	3.05
Height	13.00	7.93

In the above dimensions of the chief dam, the length of the foundation is less than that above the surface. This appears ridiculous because in this position the dyke would not have sustained the pressure of water. The reason might have been that for foundation the engineers had utilized the length-wise base of the hill. As the hill goes up its breadth decreases, therefore, to cover this gap the wall constructed above the surface was longer in length. Where the base of the hill was not available the proportion of the measurement was opposite to it.

The longest embankment wall is near the village Asotia where it is 1261.5 m. Then comes Bansol, Kankroli and the chief dyke. The total length of the water harvesting structure is 3,912 m. The breadth of the chief weir is larger than all the other parts of the structure. The measurements of three open rooms (*koshtha*) with *burj* are recorded by the author. The one adjacent to the palace is 30.5 m. in length while the exit pathway is 15.25 m. Its circumference is 45.75 m. and height 18.3 m. The size of the body in the middle is 45.75 m. in length and 22.87 m. is its exit pathway. Its circumference is 68.63 m. while the height comes to 21.35 m. The last one is equal to the first one in size.[27] Along with the geometrical designs, the awareness of the architect about the aesthetic proportion deserves appreciation. These open enclosures were meant for the rulers.

Similarly, the author gives the measurements of three rooms on the dam of the Kankroli.[28] This contains three *mandaps* of four pillars besides a *sabha mandap* (assembly hall)[29] (Figure 4.5). The *gaughat* (site for cows) is square in shape. It measures 33 m. in length and breadth and its height is 3 m.[30] Its real part is 33 m. in length while the breadth is 6.10 m. Its height is only 1.83 m.[31] An access road for an old temple of Shiva was also laid.[32]

Further, the author mentions two chambers located on the dam near the village Asotia. Its size is 14 m. in length, 8.5 m. in exit and 14.64 m. in height. One is octagonal (*ashtakoniya*) while the other is semi-circular (*ardha-chandra*). This is small in size compared to the former which is 14.64 m. in length, 6.1 m. in width and 7.32 m. in height.[33] The chamber is decorated like the half-moon, therefore denoting the abode of Shiva. Three *mandaps* on the front part of the dyke lend it beauty.

Similarly, three enclosures were erected on the weir near the village Bansol. The chamber on the corner is rectangular in size whose length and breadth are 12.2 m. while the height is 12.2 m. It is extremely beautiful. There is a Persian wheel, called *rahant* in Hindi which ensured continuous supply of water.[34]

The chamber in the middle is again semi-circular (*ardha-chandra-kar*) and equal in length and exit.[35] Its height is 10.37 m. The third

Figure 4.5: View of Chambers on the Kankroli Dam

chamber is octagonal (*ashtakoniya*) and again equal in length and breadth.[36] Its height is 5.5 m. and it is named *Kamal Burj*. At this site is a *mandap* of white marble, extremely beautiful with eight female statues (*putlikayen*).[37] There are three more *mandaps* in the front portion of the dyke.

The *Rajprashasti* gives the total number of constructions on the body of the structure. Twelve chambers (*koshtha*) were erected in different locations as well as 21 *mandaps*. In addition to it, the chief dam has two *mandaps* decorated with fine sculptures.[38]

OUTLETS OF THE HYDRAULIC STRUCTURES

The *Rajprashasti* goes further and gives the position of the three outlets, called *ota*,[39] from where the surplus water flowed out. The first sluice is 152.5 m. in length and 6.1 m. in breadth. The length and width of the second structure corresponds with the first while its height is 1.53 m. The third is the longest at 183 m., with a width of 6.1 m. and a height 1.22 m.[40] All the three outlets extend till the end covering its entire width. From the corners the water was drained out to the villages[41] for irrigation (Figure 4.6 A & B). The *Vir Vinod* mentions a fourth outlet. In case there's excess in the reservoir, the surplus flows out over the *gaughat*.[42] How did the people take water to the villages? Perhaps the peasants dug out canals from which water was conveyed to the fields through drains. The British reports and gazetters do not mention canals from the Raj Samand.[43]

EXTENT OF THE WATERBODY

Raj Samand was one of the longest dams in Mewar with an extensive span. Some of the villages were completely submerged in the waters of the dam while some partially lost their areas to it. The six villages that completely lost their identity were Siwali, Bhagwada, Bhana, Luhan, Bansol and Gudhali while the boundaries and partial area as of another six villages were sub-merged in the tank namely, Morchana, Pasund, Kheri, Chhapar-

(A)

(B)

Figure 4.6: Outlets of Raj Samand

kheri, Tasol and Mandawar.[44] All these villages are shown on map (Figure 4.7).

Not only the boundaries of the villages, but also the ponds of three villages namely, Kankroli, Luhan and Siwali, natural depressions (*niwan*), step wells and wells were subsumed in the dam. In all they total up to 30 in number.[45] This shows the enormity and extent of the dyke. Its length, as said above, comes to 3,912 m. which is higher than the established norms of the period. According to the author, it should not be more than 3,660 m., as recommended by Vishwakarma, the lord of construction.[46] The plans of entire dam and the chief marble portion would show its size and engineering skill (Figures 4.8 and 4.9).

This tank receives the water of three rivers namely, the Gomti, the Tal and the Kelwa. The testing time for the dam came when the water from the rivers gushed into it in the month of *Bhadrapad* of VS 1730/AD 1673. The flow was so forceful that a large number of houses in the area were submerged. Next month, when the water from the Gomti entered into the tank, the water level in it rose to 3.66 m. only.[47] The water was retained in the tank and served as a test of its strength and stability.

This provided an opportunity for the Maharana to celebrate constructing the great dam. He performed the prescribed religious ceremonies and gave donations to the Brahmans and others after which he went sailing in the waters of the dyke. The boats were made by the mastercraftsmen of Lahore, Gujarat and Surat, the poet Ranchhod Bhatt informs us.[48] Since there was no possibility of boating in Rajasthan, it was impossible to get local experts for making boats in the region.

MANAGEMENT OF THE CONSTRUCTION WORKS

The management and organization of the construction work of a massive dyke like Raj Samand was an essential component of the project. The evidence contained in the *Rajprashasti* attests to this. The work appears to have been organized in a very

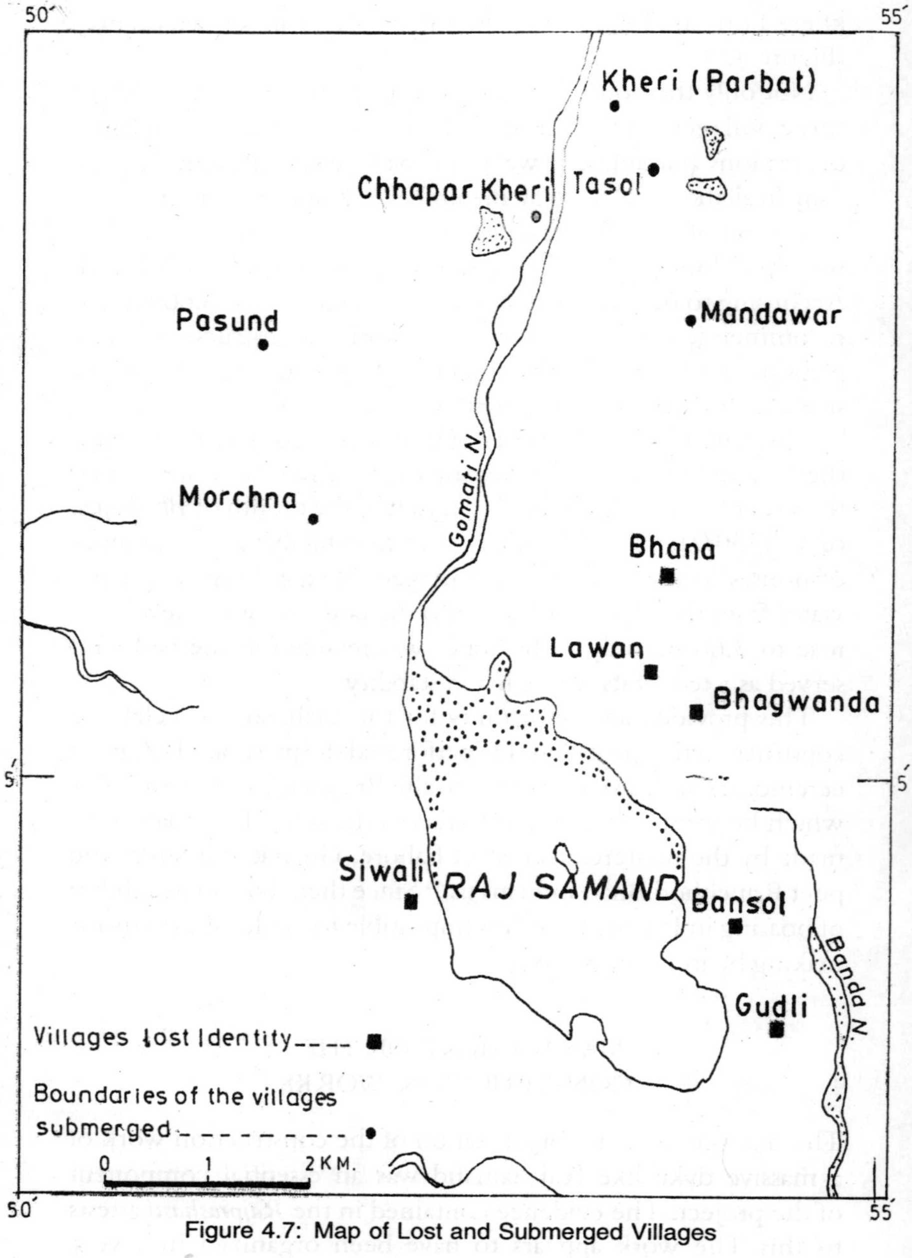

Figure 4.7: Map of Lost and Submerged Villages

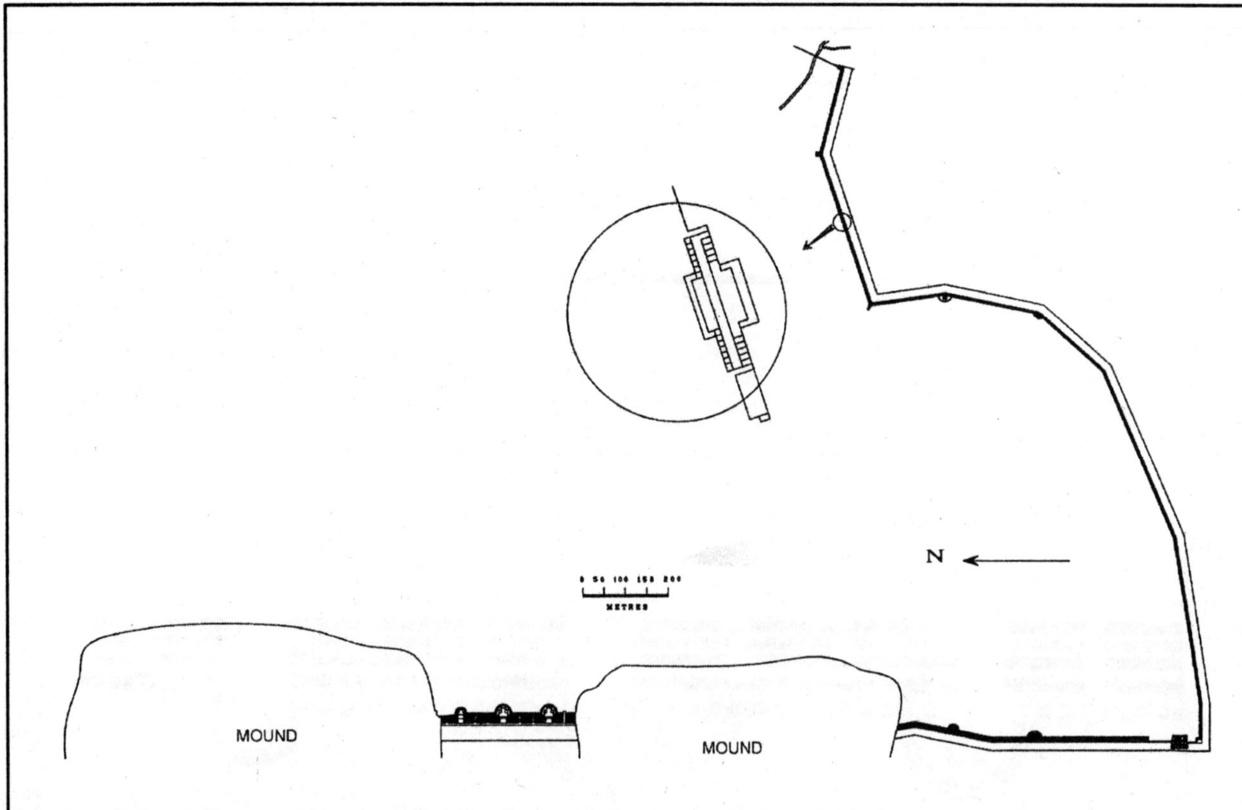

Figure 4.8: Ground Plan of Entire Raj Samand Dam

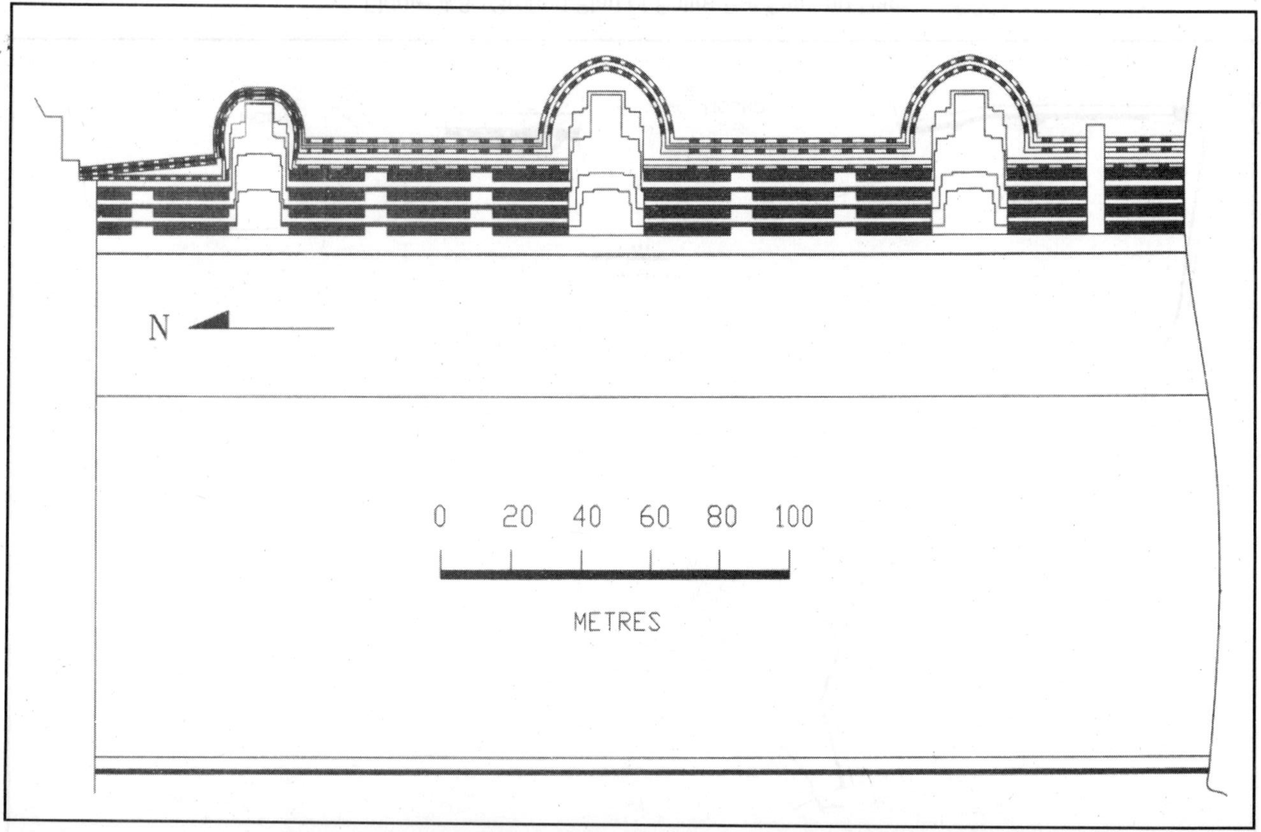

Figure 4.9: Plan of Main Raj Samand Dam, White Marble Pavilion

meticulous way. Broadly three departments were temporarily established: engineering, finance and administration. Every department seems to have been headed by an officer. Except the department of engineering, no direct information on other corresponding units came to us. For this we have to depend on indirect evidence.

The department of engineering was headed by one Mohanji, the son of Sukhji.[49] He was the chief architect and engineer (*sutradhar*) of this hydraulic project. He himself prepared the ground plan of the body with the assistance of his subordinate overseers. Before initiating the work he seems to have got the approval of his master Maharana Raj Singh. He executed his dream plan and supervised every stage of the project and ensured that basic parameters of the hydraulic norms were properly adhered to. The stages of the work demonstrate his deep knowledge about different aspects of dam building technology. The laying of the foundation on a level earth below the sub-soil water to sustain the overall weight of the superstructure was the first stage.[50] This means the chief engineer was aware of the soil erosion factor which necessitated a very deep foundation.[51]

The object of raising a wooden structure on the foundation was probably to ease the task of the masons while erecting the masonary structure over it. The probable reason behind the construction of this wooden fabric was that any change or defect could be easily rectified. This was according to the measurements shown on the draft design. These two examples are sufficient to show the expertise of the chief engineer in erecting hydraulic works such as the dam.

There was a staff of overseers (*gajdhar*, Persian *gazdhar*) who assisted the chief engineer.[52] The *Rajprashasti* records at least three *gazdhars*: Vagh, Mukund, and Kilyan, the son of Jagannath.[53] The text of the *Rajprashasti* reproduced in the *Vir Vinod* gives eight names: Vagho; Mukund; Kilyan, the son of Jagannath; Urjan, the son of Lalo; Lalo; Jaso; Harji; and Jagannath, the son of Megho Mano.[54]

EXPENDITURE ON THE DAM

Raj Samand is the first waterbody of which minutest details of expenditure incurred are available. The *Rajprashasti* records the expenditure from the beginning till the end, that is, from 12 January 1662 to June 1675.[55] At the outset, it gives the account of the expenditure spent on the first part of the work which totals Rs. 4,664,625 and 4 *annas*. Then it gives the breakdown of this expenditure as follows:[56]

	Rs.	*Annas*
Ram Singh Ranawat	2,736,497	8
Uncle of the Maharana	5,04,880	4
Mohan Singh Sisodia, son of the Maharana	2,12,538	-
Pancholi Sanwal Das	4,78,107	-
Wages for the digging of squared land (*Chaukariyas*)	32,601	4
Expenditure on the *Pratishtha* ceremony	7,00,001	-
Total	3,964,624[57]	-

This amount was spent under two heads. One was the wages of workers who were engaged in the construction of the dyke.[58] This amounted to Rs. 3,202,880 and *annas* 4; Rs. 761,744 were probably in the form of presents or salaries to the *thakurs* who were supervising the construction work.[59] The total expenditure thus came to Rs. 3,964,624.4, which is short of only 12 *annas* mentioned above. The consolidated amount spent in the second part of the dam was Rs. 10,507,608.[60] The *prashasti* records only the amount which was spent through the supervisors who were officially appointed. The Maharana is said to have spent a huge amount of wealth on gifts, presents and on various ceremonies which were not part of the regular expenditure on the waterbody. The money spent by the *jagirdars* for digging and other allied works was in addition to what is recorded.

The evidence provided by the *Rajprashasti* creates confusion. It records the detailed account of expenditure incurred in the first part of the dyke while for second part it registers only a consolidated amount for which no detail is mentioned. The evidence encourages us to infer that both the amount constitute the total expenditure on the waterbody. This may include wages of the workers, masons and craftsmen and the salaries of the supervisors. While Shyamal Das, the author of the *Vir Vinod*, writes that the amount mentioned for the first part is said to have been spent on the work on the tank, the latter figures stated for the second part were incurred on special occasions, probably for giving gifts, donations and celebrations.[61] Shyamal Das' assumption does not hold much water because the *Rajprashasti* specifically mentions that the money spent on the above items was not included in the expenditure on the weir.[62] Thus, we may conclude that the total amount incurred on the Raj Samand came to Rs. 14,472,232 and 4 *annas*.

CONCLUSION

The study of the Raj Samand shows the aesthetic sense and mathematical precision of the architects and engineers, using which they built this mammoth yet marvellous structure which has remained standing for the last three and half centuries, facing all kinds of natural calamities. This structure was the outcome of the accumulated experience of master builders in constructional fields. The Raj Samand is thus an excellent example of indigenous dam technology. Besides this, the platforms have beautiful carvings which make the structure unique and magnificent, a fine example of expert planning and workmanship.[63]

NOTES

1. This epic was initiated on the *Magh Badi* 7, VS 1718/AD 1661 on the same day of the construction of the dam which means the poet was a witness to the entire process of the building of the structure. It appears to have been closed in VS 1738/AD 1681, covering the period

from 1661 to 1681. It was inscribed on the order of Maharana Jai Singh in VS 1744/AD 1687.

2. The survey was conducted with the financial assistance of the Centre of Advanced Study, Department of History, AMU, Aligarh. I express my gratitude to Ghulam Mujtaba (photographer) and Zameer Ahmad (technical assistant) for their help. They took detailed measurements and photographs of the structure.
3. Cf. *Rajprashasti*, pp. 90–1. The names of the 16 villages where Raj Singh found suitable location for the construction of the weir were: Dhoyanda, Sanwad, Siwali, Bhagawada, Morchana, Pasund, Kheri, Chhaparkheri, Tasol, Mandawar, Bhan, Luhana, Bansol, Gundhali, Kankroli and Madha.
4. Ibid., pp. 91–2.
5. Ibid. The literal meaning of the term *setu* is bridge but the author uses it for a dam. The word *tadag* or *tarag* has Sanskrit origin and means tank. The author writes that the Maharana decided to construct a tank (cf. ibid., pp. 91–2).
6. Kavi Maan, *Raj Vilas*, ed. Moti Lal Menariya, Varanasi: Nagripracharani Sabha, VS 2015, pp. 93–5. It is a historical epic in Braj-Rajasthani. Maan was a contemporary of Maharana Raj Singh and composed this work covering every aspect of his patron. He gives a pathetic description of the famine of VS 1717/AD 1660 which devastated human beings, birds and animals alike. Thousands of men and women abandoned their kith and kin and parents sold off their children. These conditions are attested to by a Jain traveller who writes that the scarcity in Mewar continued from VS 1715/AD 1658 to VS 1717/AD 1660 (cf. Yati Jai Chand, *Saiki*, ed. Muni Kanti Sagar, Jaipur: Chhutan Lal Bairathi, n.d., pp. 11–12).
7. *Vir Vinod*, vol. II, pt. I, p. 444.
8. Ibid., pp. 444–6. The stories of the murders of a *Barath* (bard), the elder son Sultan Singh, wife and a *purohit* are recorded in the *Vir Vinod*. On this count the Maharana enquired from the Brahmans about the way of repentance for these heinous crimes. In turn they suggested three ways: to immolate under a dry *peepal* tree; to construct a big tank; be killed in battlefield. The Sisodia chief accepted the last two options (cf. ibid., pp. 44–6).
9. *Rajprashasti*, pp. 444–6.
10. Ibid., p. 115.

11. Ibid., p. 89; *Vir Vinod*, II, pt. I, p. 448.
12. The author equates the involvement of workforce in digging the site of the proposed dam with a mythological story. He says that as the 60,000 strong sons of Sagar dug out the land for the construction of the sea, similarly millions had started excavation for the construction of the second sea (cf. *Rajprashasti*, p. 93).
13. Ibid.
14. Ibid., p. 94.
15. The terms used for the water lifting machine are *ghatiyantra, jalyantra* and *chakrabharta*. Oxen were employed to draw water from the deep pits or wells dug out to reach the rocks. So undoubtedly these were Persian wheels (cf. ibid., pp. 94–5).
16. The drained water was carried by the people by digging canals to their villages and thus villages were transformed into cities (cf. *Rajprashasti*, p. 96). Here the author meant that agriculture flourished which brought prosperity.
17. The depth of the foundation gets special praise from the author. He compares this with that of the mythological bridge constructed by Lord Rama. He writes that the land was dug to reach the underground water. Here, at this level, multi-coloured fishes were found. Since this process was not followed by Lord Rama, it was destroyed very soon (cf. *Rajprashasti*, p. 100).
18. Ibid., p. 97; *Vir Vinod* records the year VS 1722 which appears a slip (cf. ibid., p. 448).
19. The poet praises the Maharana for blocking multiple sources of water with stone slabs soaked in lime-mortar (cf. ibid., p. 98). Then he compares him with Rama and Krishna. As Rama vanquished Kabandh, a demon (*rakshas*), and succeeded in controlling the power of water, similarly, Krishna protected the people from the anger of Lord Indra by holding aloft the Govardhan mountain on his finger. Similarly the Sisodia chief protected the people for seven years by holding the weight of entire earth (cf. ibid., p. 99).
20. Rajiv Sharma defines the term *sharibandh* as back wall which does not seem correct. See 'Raj Samand Dam', op. cit., pp. 219–23.
21. The poet praises the Maharana while comparing his efforts to those of Lord Rama. The dam constructed by the former to cross the sea to reach Lanka could not last because neither was water removed nor was it built on earth, while the Maharana raised the dyke after removing water and digging a foundation. It was built

by skilled labour using stone and mortar. This, in fact, is in praise of the engineers of the waterbody (cf. *Rajprashasti*, p. 100).

22. Ibid., pp. 104–5.
23. Ibid., pp. 108–9.
24. Ibid., p. 110.
25. Ibid., p. 103.
26. Ibid., pp. 112–23. The entire eleventh chapter and the first four *shlokas* of the twelfth chapter provide information on measurements of each segment along with the three outlets.
27. Ibid., *sarg* 11, *shlokas* 7–10, pp. 113–14.
28. The room towards the *sabha mandap* is 17 m. in length while the exit is 8.5 m. long. Its height comes to 22 m. The length of the middle structure is 22 m. and the exit is 9 m. while the height is 23 m. The room in the eastern direction is 17 m. in breadth and a 9 m. exit. The height comes to 22.6 m., cf. ibid., pp. 117–18.
29. Ibid., p. 119.
30. Ibid.
31. Ibid.
32. Ibid., pp. 118–19.
33. Ibid., p. 120.
34. Ibid., p. 121.
35. The length and size of the exit are 7.3 m. cf. ibid., p. 121.
36. The length and width are 18 m., cf. ibid., p. 122.
37. Ibid., p. 122.
38. Ibid., p. 126.
39. The term *ota* means a place from where excess or surplus water of the tank is drained out. This is also called *pariwah* (flowout path), *chadar* (overflow) and *diwar* (full flow).
40. Ibid., p. 123.
41. Ibid., pp. 123–4.
42. *Vir Vinod*, II, pt. I, p. 449.
43. The gazettees compiled by the British authorities mention that this reservoir of Raj Samand did not have any means of distributing the stored water. It is reported that between 1884 and 1886 canals were constructed through which about 2,000 acres were irrigated annually (cf. Erskine, op. cit., vol. IIA, p. 9). This statement is negated by the evidence of the *Rajprashasti* which clearly indicates the existence of three outlets. The water flowed from these outlets to

be used by the villagers through unlined canals (cf. *Rajprashasti*, p. 124).

44. Ibid., p. 124.
45. Ibid., pp. 124–5.
46. Ibid., p. 125.
47. Ibid., pp. 128–9.
48. Ibid., pp. 130–1.
49. Ibid., p. 264. *Vir Vinod* reproduces the total text of the *Rajprashasti* but the name of the *sutradhar* is surprisingly missing (cf. II, pt. I, p. 634).
50. Ibid., p. 100.
51. This is evident from a stanza where the poet describes the destruction of the dam, said to have been built by Lord Rama. It fell because the foundation was not laid down after draining the water, writes the poet to highlight the achievement of his patron. But this mistake, he says, was not repeated in building the Raj Samand Dam because the work was in the hands of expert architects. This is the reason for its strength (cf. ibid. This point is highlighted by Rajeev Sharma in his article, Raj Samand Dam, op. cit., p. 220).
52. For the meaning of *gaz* see H.H. Wilson, *Glossary of Judicial and Revenue Terms*, Delhi: Munshiram Manoharlal, 2nd edn., 1968, p. 171. For the term *gajdhar* see *Rajasthani Sabad Kosh, I Rajasthani*, ed. Sitaram Lalas, p. 669. B.P. Sakariya and Bhupatiram Sakariya, (eds.), *Rajasthani Hindi Shabdkosh*, Jaipur: Panchsheel Prakashan, 1977, p. 303.
53. *Rajprashasti*, p. 264.
54. *Vir Vinod*, II, pt. I, p. 634.
55. On *Magh Krishna* 7, VS 1718 the work on the dam was commenced; in the month of *Asadh*, VS 1735 the last ceremony of its completion was performed (cf. *Rajprashasti*, p. 222); *Vir Vinod*, II, pt. I, p. 451.
56. Ibid.
57. The total given in the *Vir Vinod* is different though the margin of difference is negligible—only 4 *annas*, *Vir Vinod*, p. 451.
58. *Rajprashasti*, *sarg* 21, *shlok* 12, 15, p. 224, *Vir Vinod* gives a total of Rs. 3,964,623.12 which is 4 *annas* short of the figures of the *Rajprashasti*. The breakdown given in it appears to be slip. Instead of 3,202,880 *annas* 4, it records Rs. 32,002,880 *annas* 4 (cf. II,

pt. I, p. 451). Further details given by Shyamal Das are different. According to him, Rs. 3,202,880 were spent on the filling of the foundation and construction with lime-mortar and the remaining Rs. 761,743.8 on the cutting and finishing of stones.

59. *Rajprashasti, sarg* 21, *shlok* 22, p. 226. The *Vir Vinod* gives the total as Rs. 10,547,584 which does not correspond with the *Prashasti.* It further gives the breakdown of the expenses. According to it Rs. 3,964,623 and *annas* 12 were spent on the work of the tank while the remaining Rs. 6,582,960 and *annas* 4 were for gifts, donations and ceremonies.
60. *Vir Vinod,* II, pt. I, p. 451 n. 1.
61. *Rajprashasti, sarg* 21, *shlok* 23–5.
62. The total in the *Vir Vinod* is recorded as Rs. 14,472,207 and 12 *annas* which is short of about Rs. 25 from the actual total. It is not known from where the author of *Vir Vinod* has taken the amount of Rs. 10,507,584 (cf. ibid., p. 451) while the *Rajprashasti* clearly mentions Rs. 10,507,628 (cf. *shlok* 22, p. 226).
63. For sculptural study, see Neelima Vashishtha, 'A Note on the Sculptors of the Raj Samudra Lake in Rajasthan', in Ahsan Jan Qaisar and Som Prakash Verma (eds.), *Art and Architecture,* Felicitation Volume in Honour of Professor S. Nurul Hasan, Jaipur: Publication Scheme, 1993, pp. 111–17.

CHAPTER 5

Jana Sagar or *Bari (Badi) Ka Talab*

The creation of a tank near the city of Udaipur demonstrates two objects of Maharana Raj Singh: one, to provide continuous water supply to the people of city of Udaipur and second, irrigating the fields of the surrounding villages. The tank was located 9.5 km. north-west of the capital and was known by two names, viz:, *Bari (Badi) ka Talab* and Jana Sagar. The former is the name of the village where the tank is located while the latter designation came into vogue after the name of Janade, the deceased mother of Raj Singh.[2] The foundation of the tank was laid on 15 November 1664 (*Margashirsha, Vadi*, 10, VS 1721) and the construction was accomplished on *Baisakh Shukla* 3, in VS 1735/AD 1678.[3] It took more than 13 years and 6 months to complete the work. It is to note that the Maharana initiated two irrigation projects within a span of six months: the foundation of the Raj Samand on *Baisakh Shukla* 13, 1664 (in the month of May 1664)[4] while work on the Jana Sagar or *Bari (Badi) ka Talab* began in the November of 1664.

The site selected for the dam shows the wisdom of the civil engineers of Mewar. The three sides of the area are surrounded by hills with one side between two hills being open. The declivity of the vast plain of the foothills is a saucer-shaped basin spread over an area of 15.36 sq. km. Prior to the construction of the embankment, water used to flow down near the capital city of Udaipur. The village Bari/Badi[5] is located on the other side of the mountains and a lower end of the hill is used as a passage to reach the village.

The western side of the dam appears as a web of mountains, which is the source of water. In fact, water comes from three directions with great force, causing cuts in the hills. In local parlance the water flow from the surrounding hills located in three directions is called the Ubeswar River[6] (Figure 5.1).

The so-called river flows through a gorge originating from the hills. This overflow was dammed by the erection of an embankment wall between the two hills. The height of the wall is equal in size to the hill located in the eastern direction which also serves as a pathway to village Bari Uparli,[7] that is, the village located on the upper part of the hills.

The straight wall is projected towards the tank which is a five-storeyed structure. Two lateral staircases at two opposite ends were constructed to reach from the ground to the first floor. Nine stairs in each direction lead to this floor. Then eight stairs on each side lead to the second and the third platforms. Again nine stairs help to reach at the fourth floor which is the top platform. In all, 42 steps in one and 84 in both the directions were constructed between the ground and the top platform. On the top, every *chauki* has a niche located in the middle of the wall facing the tank. This is evident from the front view of dam (Figure 5.2).

In the middle of the embankment wall two *chhatris*, one above the other, were built and extended towards the tank. On the upper cenotaph, an inscription inscribed on the slab gives details about the dam. The stone document assigns name to the tank as Jyan Sagar (Jana Sagar), after the name of Maharana Raj Singh's mother.[8]

The outlet or channel is located on the south-east of the embankment wall. Three outlets at different levels were cut out in the wall. The first spillway is at the tank's ground level while the second and third are at 3.66 and 6.71 m. above the ground, respectively. On the third floor, there are bridges (like sluices) erected through which water moves into the channel. The edifice is about 9.45 m. above the ground level. The sluices are arch-shaped. At this spot the south-east projected platform is constructed to provide support to the canal which runs along the eastern hill (Figure 5.3).

The back wall is considered an important component of the dam, it is an additional support wall to strengthen the dam. The support wall of this structure is in the form of nine projected

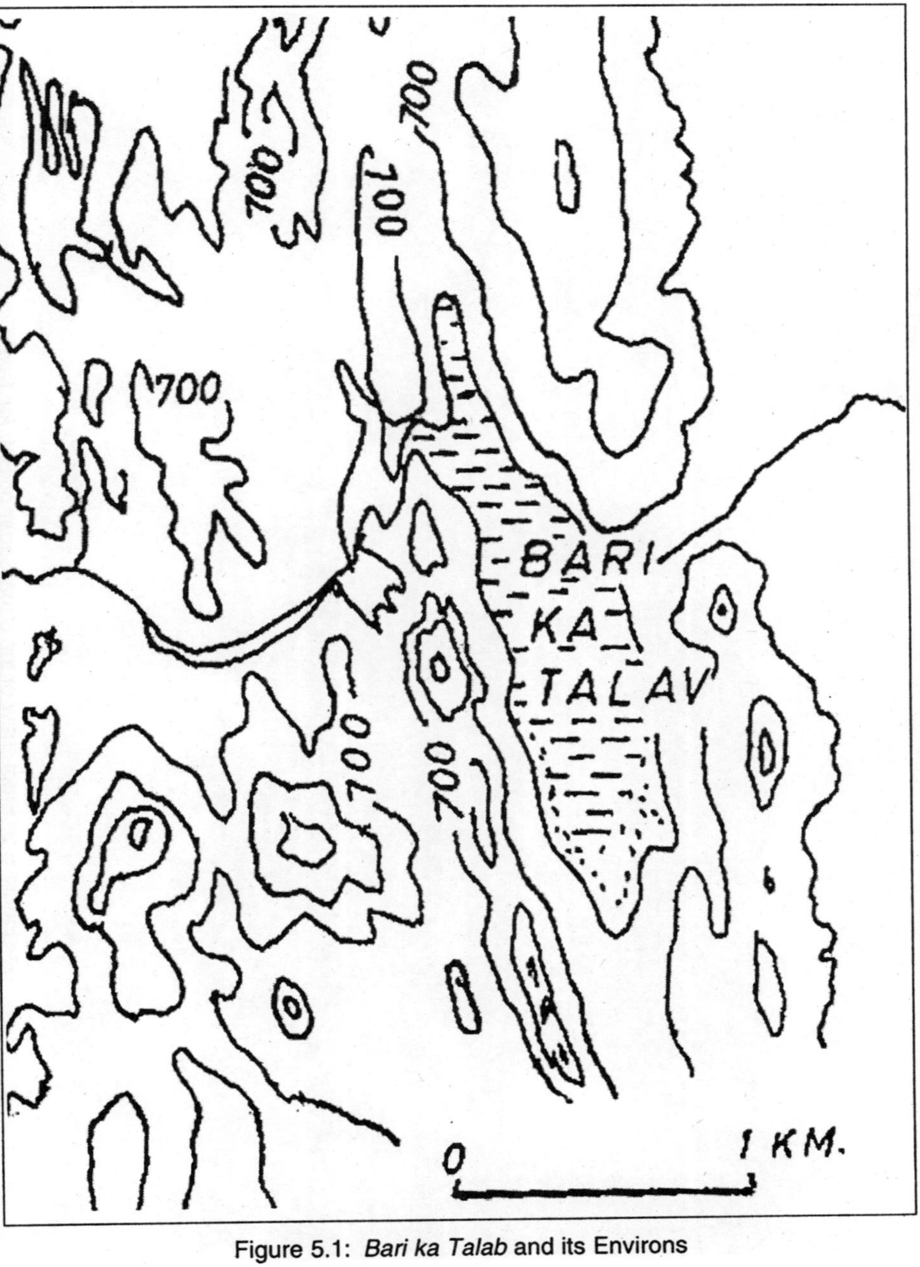

Figure 5.1: *Bari ka Talab* and its Environs

Figure 5.2: Front View of the Jana Sagar or *Bari ka Talab*

Figure 5.3: Outlet of Jana Sagar

platforms. Support is provided stage by stage to give strength to this wall which would also in turn support the embankment wall. The breadth of each platform increases as it goes down. In fact, it is like a slope (Figure 5.4). Two motives appear to have been active in the minds of the engineers: first, in case of water flowing over the barrage wall due to torrential rains, it would flow down over the back wall and secondly, the platforms would protect the back wall itself from any transit tensions. In case of overflow from above the embankment wall to lessen violent movement of water, the back wall was raised in the form of descending order so that force of overflow would reduce on every fall on platform and thus could not damage the retaining wall. At last, when water reaches the last platform its force would automatically be minimized.

Keeping this in mind, the engineers kept the breadth of the last projection bigger than others. This pyramid-like design appears to be a unique feature of this dam.

DIMENSIONS OF THE DAM

The inscription on the stone slab of the *chhatri* records the measurements of the dyke in *gaz*.[9] It gives the height (*ude*) of the structure as 31 m. the length and breadth as 183 and 55.5 m. respectively.[10] The *Janasagar Prashasti* does not gives measurements but eulogizes the builder of the dyke and his family.[11] The measurements discussed below have been obtained by physical survey of the structure.[12] The embankment wall runs north-west and south-east. Its length between the two hills comes to 192 m. while its breadth is 63 m. The retaining wall is erected at the distance of 32 m. and this entire space was stuffed with earth and lime mortar. The breadth of the back wall measures 2.5 m., the total breadth (including the embankment wall) thus coming to 97.5 m. This provided strong support to the barrage wall (*pal*) which could sustain the violent force of the water. The size of the outlet comes to 8.55 m. in length. Both the platforms are rectangular in shape. The first platform is 2.70 × 2.30 m. while the second is 1.50 × 0.75 m. (Figure 5.5).

Figure 5.4: View of Back Wall

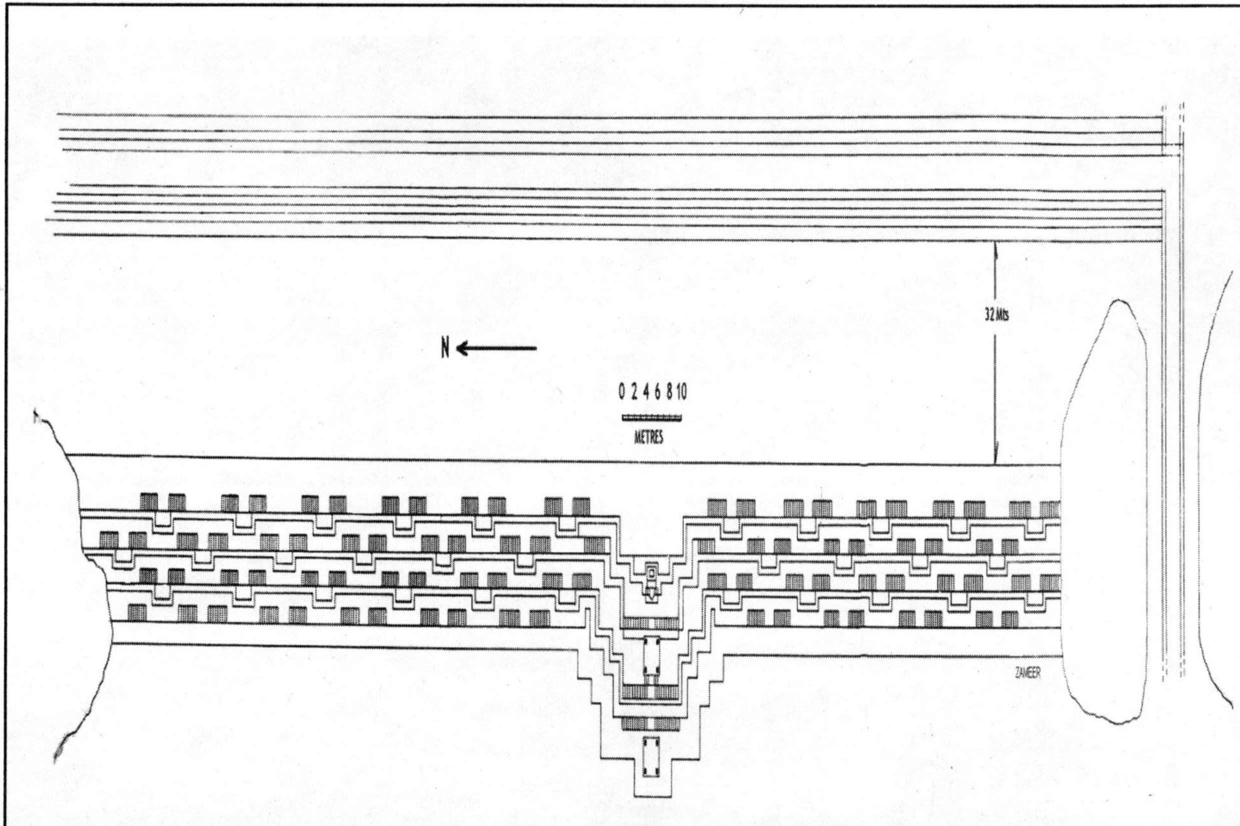

Figure 5.5: Ground Plan of *Bari ka Talab*

The comparison between the measurements recorded in the inscription and obtained through physical survey of the structure yields interesting results. To make a comparison possible, both the figures were converted to feet. Since the *gaz* in Mewar is of 0.61 m., therefore conversion becomes essential. The conversion of both the figures in feet is as follows:

	Measurement in the Inscription		Measurements as per the Survey	
	Gaz	Feet	Metre	Feet
	I	II (× 2)	III	IV
Length	300	600	192	630
Breadth	91	182	63	206

The difference in length is 9 m. and 7.3 m. in breadth. Two reasons may be assigned for this gap: first, either an error in fixing the size of the *gaz* of the seventeenth century in modern times or the difference in fixing the point or spot at both the corners from where the measurements were taken by us. Nevertheless, the gap in measurement is not very alarming and in fact quite close to each other.

The height of the dyke is recorded as 31 m. in the inscription but the technical description of recent period gives 11.6 m. as the maximum water level. Here the difference of 19.4 m. comes to surface. The simple reason for this could be that silting must have reduced the depth of the tank in recent times. The engineers measured the height of the dam above the foundation when it was first constructed and the surface of the tank was free from any silting. Therefore, we may not be far wrong if we deduce that during the last 300 years silting has raised the level of the tank.

COST OF THE STRUCTURE

The expenditure incurred on the construction of the weir is neither recorded in the inscription nor by the *prashasti*. The latter mentions only an amount of Rs. 261,000 which appears to have been spent on the ceremony at the completion of the dyke.[13] The writer of the *Rajprashasti* records a consolidated amount of Rs. 680,000 incurred on the construction of the reservoir[14] but it does not offer a breakdown of the expenditure, as in the case of the Raj Samand.

CIVIL ENGINEERS

The ground plan and the construction of the dyke was carried out under the supervision of Kisna, the son of Jasa. He was the *gajdhar-sutradhar*[15] and appears to have had the sole charge of the engineering aspect. He had prepared the ground plan and supervised its entire construction right from the laying of the foundation till its completion. The *prashasti* gives additional information about the officer-in-charge of the construction (*kamthana*). It records that the foundation of the structure was dug out by Sangram, son of Nathu. It becomes clear from the text that he was the overseer or supervisor of the construction work (*kamthana ra gajdhar suttar*).[16] Reading of the *Jana Sagar Prashasti* and inscriptions on dam makes it clear that Kisna was the planner of the project while Sagra was its actual executor under whom construction was undertaken. Thus these two, experts in their respective roles, were responsible for the technical aspect of the dyke.

The inscription on the dam records the names of three persons: Sah Dayal, Sah Ramchand and Sah Rupsi. Their names were associated with the *pal* (embankment wall) which means that either a specific portion of the dyke or the entire construction of the structure was carried out under their supervision.[17] They also looked after the financial part of the project. We do not get any evidence of the salary of the engineer or the wages of the diggers or labourers. What is clear is that the work of the waterbody was jointly looked after by separate teams of accountants and engineers. This hints at the level of organization of the irrigation works.

RELIGIOUS GRANT ON COMPLETION OF THE WATERBODY

The successful completion of the waterbody was considered an auspicious occasion. In fact, it was an occasion for celebration, both for the engineers, managers and the promoter or the spirit behind it. It was a great success for the civil engineering technology of the medieval times. It was also a day of celebration for the ruler who organized a *pratishtha* ceremony and performed *tuladan* and granted villages to the Brahmans on the occasion. Two villages, Galundchitor and Devpur,[18] were given to Purohit Garib Das in charity. This *pratishtha* ceremony became a symbol of pomp and show in which a large number of guests from neighbouring states were invited in addition to the local nobles. Large donations both in cash and kind were given on the occasion. This became the integral part of the ethos of the rulers.

NOTES

1. *Vir Vinod*, vol. II, pt. I, p. 448.
2. It is mentioned that the Maharana named the newly-constructed tank as 'Jana Sagar' and dedicated it to his deceased mother, Janade, cf. *shloks* 47 and 50, *sarg* 8, *Rajprashasti*, op. cit., p. 88.
3. There is an inscription in Mewari on a stone slab of the tank and put on the wall of the cenotaph facing the reservoir. It records the date *Mangsar* 10, VS 1721/AD 1664 as the ceremony of the foundation of the tank and the *Baisakh Sudi* 3, VS 1735 as the day of its completion. The confusion arises due to the date mentioned by Ranchhod Bhatt in his book where he records *Magh Shukla* 10, VS 1725 as the date for the foundation of the tank (*tarag*) cf. *Rajprashasti*, op. cit., *shlok* 47, *sarg* 8, p. 88. This date does not correspond with the foundation or the completion date of the structure given in the inscription. Bhatt further records that the Maharana made a *tuladan* of silver on this occasion, a kind of donation in which silver or gold or any item of equal weight of the donor is distributed among the needy. Cf. ibid.

4. Rajiv Sharma gives the year January 1664 as the date for laying the foundation of the Raj Samand Dam, which does not seem correct, Raj Samand Dam: *PIHC,* 1993, p. 220. He appears to have mistaken the date of *muhurt,* that is, worship of site as the actual beginning of the work on the dam. The *Magh Krishna* 7, Wednesday VS 1718 was the date declared as auspicious by the astrologers (*Rajprashasti*, op. cit., p. 92) but the actual digging of the foundation was commenced on *Baisakh Shukla* 13, VS 1721 (ibid., p. 97).
5. The technical report mentions two villages of the same name: (1) Bari Uparli, and (2) Chhoti Bari.
6. This is reported by a person named Ram Lal of village Bari on 22 September 1999 when the author surveyed the embankment for the second time. It is interesting to note that nowhere do we come across any reference about this river. There is a possibility that the name of the river was coined by the local populace. The word Ubeswar could be identified with Shiva. The technical report depicted on the signboard at the site (in Hindi) calls it the basin of the river Banas. Ubeswar seems to be a seasonal tributary of Banas.
7. The village lies within the *tahsil* of Girwa, district Udaipur.
8. Janade was the mother of the Maharana and wife of the deceased Maharana Jagat Singh. She was the daughter of Rathor Raj Singh Mertiya. Janade died before the ceremony to mark the completion, so the tank was dedicated to her memory, *Rajprashasti*, p. 88; *Vir Vinod,* vol. II, pt. I, p. 448.
9. In Mewar, 1 *gaz* equalled 2 feet which is called *Silawati* (cf. *Vir Vinod*, I, p. 212).
10. For the original text see, Appendix 2, *Rajprashasti*, pp. 272–7.
11. This *prashasti* is different from the inscription on the dyke. It was written by Shri Laxminath Bhatt and Bhaskar Bhatt. They were Tailang Brahmans of the village Kathore and sons of Shri Krishna Bhatt. This *prashasti* was composed in VS 1734/AD 1677 and probably inscribed in VS 1735/AD 1678. Both the inscription and the *prashasti* record the dates of the foundation and the completion of the dyke but do not agree on the latter date. The inscription does not give the exact day of the foundation of the *pal* but records only the year VS 1721/AD 1664, which matches with the *prashasti*. But both the documents mention divergent years for the completion of the structure. The former records the year VS 1729/AD 1672 as

the completion of the *pal* while the latter gives VS 1735/AD 1678. This difference in dates could only be explained by the fact that the year VS 1729/AD 1672 in the inscription indicates the finishing of the work of the embankment wall while the year recorded in the *prashasti* represents the completion of all kinds of work related to the structure. This may be the reason that the latter does not use the term *pal* but uses the expression 'work completed' (cf. Appendix 2, *Rajprashasti*, op. cit., pp. 272–7).

12. I surveyed this structure along with technical assistant, Zameer Ahmad and photographer Ghulam Mujtaba of the archaeological section of the Department of History, AMU, Aligarh. I express my gratitude to them for their valuable cooperation.
13. Appendix 2, *Rajprashasti*, p. 277.
14. Ibid., p. 88. Ojha while quoting the *Rajprashasti* and *Jana Sagar Prashasti* mentions Rs. 688,000 as the expenditure, which is correct (cf. *Udaipur Rajya ka Itihas*, II, p. 885). The translator of the *prashasti* had misread the figure. See also, Ishwar Singh Ranawat, *Rajasthan Ke Jal-Sansadhan,* Udaipur: Chirag Prakashan, 2004, p. 102.
15. The term *gajdhar* means a person who possesses a *gaz*, a unit of measurement, in other words, an expert preparing a ground plan for any kind of building according to the exact mathematical calculation in terms of inch, feet and *gaz*. When the term prefixes with *sutradhar* it probably means a person who has the qualification of a ground planner and also the ability to execute the plan. The literal meaning of the term *sutradhar* is key figure, i.e. a person around whom everything revolves. He is like the keystone in the arch.
16. Appendix 2, *Rajprashasti,* op. cit., p. 277.
17. These three names are found in the inscription inscribed on the *chhatri* of the dam.
18. *Prashasti* in *Rajprashasti*, p. 277. The *Rajprashasti* records the names of the villages as Gunhanda and Devpura (cf. p. 88).

CHAPTER 6

Jai Samand Dam

Maharana Jai Singh (1680–98) followed his illustrious father in the hydraulic domain and went a step further. An anecdote popular among the people tells that when his father Maharana Raj Singh was busy with the completion ceremony of the Raj Samand in February 1676, the young prince[1] commented about its small size. The elder Maharana retorted that he should say this after constructing a bigger dam. The prince took it as a challenge.

After his accession to the throne of Mewar in 1680, Maharana Jai Singh decided to build a bigger embankment than the Raj Samand. He immediately started search for an appropriate site. Instructions in this regard were also issued to civil engineers and other experts who brought to light two sites near the capital city Udaipur. After getting approval of the new Maharana, the construction of two barrages was started.

DEWALI KA TALAB

The Maharana initiated the project of making a large reservoir tank in the village of Dewali. It is located at a distance of 2.4 km. to the north of Udaipur. Its embankment wall is long extending from the Motimahal to the mountain of the Neemach Mata temple.[2] Ojha identifies its location as that of the present-day Fateh Sagar. He states that since the height of the barrage was short and the flow of water was deficient, it could not supply water in the southern direction. Therefore, it was erected afresh by Maharana Fateh Singh in 1889 and designated as Fateh Sagar after his own name. It was strong and the wall high.[3]

Ojha's identification of the Dewali tank with that of the Fateh Sagar points to three things: first, selection of perfect site for the lake by the engineers; second, survival of the structure for more than 200 years since its completion in VS 1744/AD 1687, and third, identical catchment area and source of water for both the Dewali and the Fateh Sagar. The latter barrage has a very small catchment area of its own fed by three other tanks namely, Jana Sagar or *Bari ka Talab*, *Chhota* and *Bada Madar*. But the catchment area of the Dewali dyke appears to have been in the vicinity of the hills located in nearby areas and also the water which it could receive from the overflow of the Jana Sagar. Similar overflow from the Pichhola might have fed the said reservoir. The dyke of Jana Sagar came into existence in 1678 during the time of Maharana Raj Singh specifically before the construction of Dewali lake. We do not know about the period of construction of both the ponds at Madar but they did not exist before Maharana Jai Singh. Perhaps water streams from the hills fed the Dewali tank. This shows that the dam was as long as the present Fateh Sagar but it was renovated in later times. As the Fateh Sagar feeds the Udai Sagar through the river Ahar, the Dewali dyke too certainly performed the same function.

THUR KA TALAB

The second reservoir was constructed by the Maharana in village Thur, 10 km. north of Udaipur.[4] It is reported that it broke down but[5] the period is not specified.

Much information is not available about this pond. The Thur tank was located between the villages Dewali and Madar. There is a probability that the Dewali and Thur tanks were interlinked, the water overflow of the latter feeding the former dyke. This conjecture is based on the location of both the tanks. The catchment area of the Thur tank might have been large, the hills beyond the village Madar being the main source of its water.

It appears that the Maharana was intending to interlink all the tanks in alignment of each other in the northern direction of Udaipur. Both were connected with the Pichhola Lake located in their west. The Pichhola and Thur lakes were feeding the reservoir at Dewali. In turn, its overflow merged with the Udai Sagar several kilometres away from the city. This shows that the Maharana built a web of lakes in and around the city of Udaipur.

Besides irrigation, the Maharana was deeply concerned about the expanding population of the city which necessitated proper planning to create a permanent source of drinking water. Trading and commercial activities had brought prosperity to the city. There was a flourishing market with a large number of shops on both the sides of the road, which is described in many city-oriented *ghazals* composed by Jain monks.[6] Nainsi's house-count of the city in the 1660s was 20,060[7] which means the population may have been around one lakh. The construction of a network of water bodies for the ever-expanding population was no doubt a challenging task. Jai Singh must have realized this problem and, therefore, he tried to link the various tanks in the vicinity of Udaipur. The efforts of the Sisodia rulers prior to him and his own efforts might have raised the water level of the entire area, which is reflected in the large number of wells and stepwells in the surrounding areas of the lakes. These reservoirs were the only source of drinking water.

JAI SAMAND

The fashioning of the two tanks did not match the grand design nurtured by the Maharana. He therefore endeavoured to select a new site with full vigour. Ultimately, a site was found 48 km to the south-east of Udaipur and it was decided to erect a dam at this point. The year 1685 marks a watershed in the history of hydrology of Mewar when foundation of one of the largest dams in India, perhaps in Asia, was laid down. On the basis of survey a ground plan of dam is drawn (Figure 6.1).

Catchment Area and Sources of Water

The contemporary sources do not offer full information about the catchment area and sources of water for the dam. The *prashasti* mentions only the name of the river Gomti which was dammed

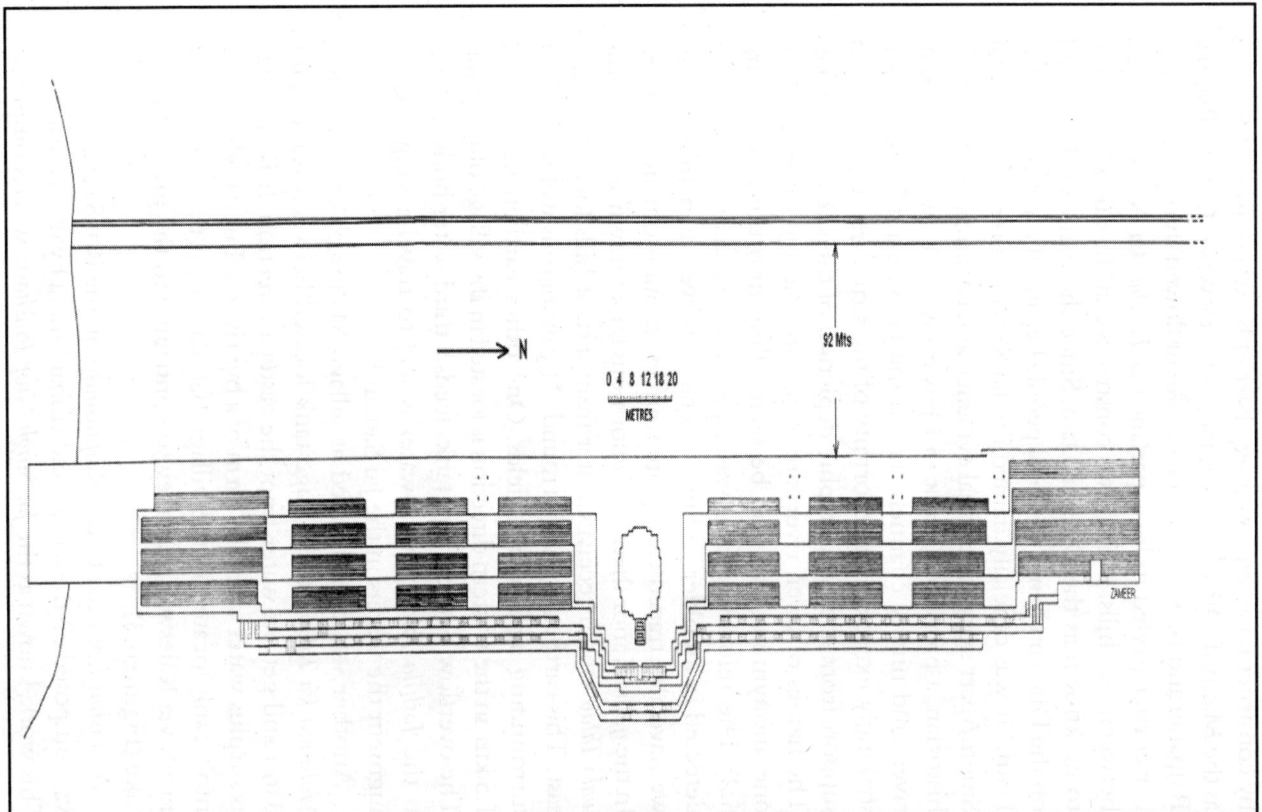

Figure 6.1: Ground Plan of Jai Samand Dam

by constructing an embankment barrier between two hills known as the Maniyal. Ojha, however, names four rivers—Gomti, Jhamri, Ruparail and Bagaar—whose confluent flowed through the Dhebar corner (styled in local parlance as *Dhebar ka Naka*),[8] located between two hills. A dam was constructed at the site which came to be known as the Dhebar Dam. Since the dam was fashioned on the Dhebar corner the local people began calling it the Dhebar Dam.[9] It was officially named the Jai Samand after its builder Jai Singh. Apart from the textual evidence and testimony of a modern historian, a proverb current in Mewar says that '(water of) nine rivers and ninety channels (fall in the Jai Samand)'. This saying strikingly expresses the enormity of the structure. This also gets support from the cartographic depiction of the site of the dyke. The names of seven rivers are traced and plotted on map while one anonymous river can be seen falling into the lake (Figure 6.2). The remaining one was either non-existent or might be detected in an older and more detailed survey map. In addition, we have also traced two more sources through our field survey. In the *qasba* Kanor there is an enormous tank known as the *Kamalwala Talab* which receives water from distant hills located on the east. This earthen tank is surrounded by numerous wells irrigating surrounding agricultural fields. One other earthen tank about 1.5 km. in the western direction is located in the village of Rajpura. The overflow of these two tanks feeds a third water body known as the *Joshila Talab* whose water is said to travel a long way to augment the waters in the Jai Samand.

Another source is located at village Mahurda, known as the *Mahurda ka Talab*. It is a huge tank located between Kanor and Bansi and gets its water from the eastern direction. It is said that its surplus water takes the form of a big river at Lunda which falls into a tank located in the village Bahida. From there the water travels via Kaleswar Mahadev to contribute to the great Dhebar Lake (Figure 6.3).

A *prashasti*, similar to the *Rajprashasti* on the Raj Samand Dam, was composed on the Jai Samand embankment by the same author. This work, known as the *Jai Singh Nrap Prashasti*, was recomposed

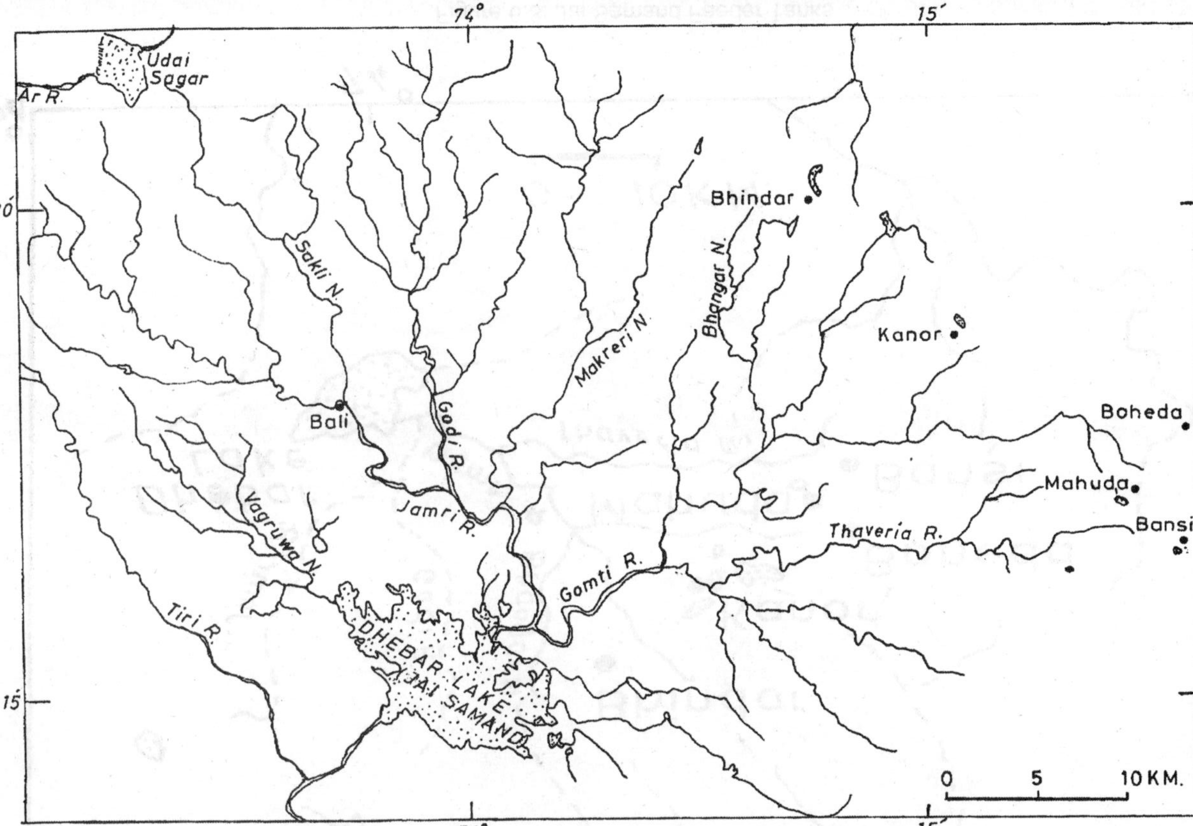

Figure 6.2: Map Showing of Sources of Water of Jai Samand Dam

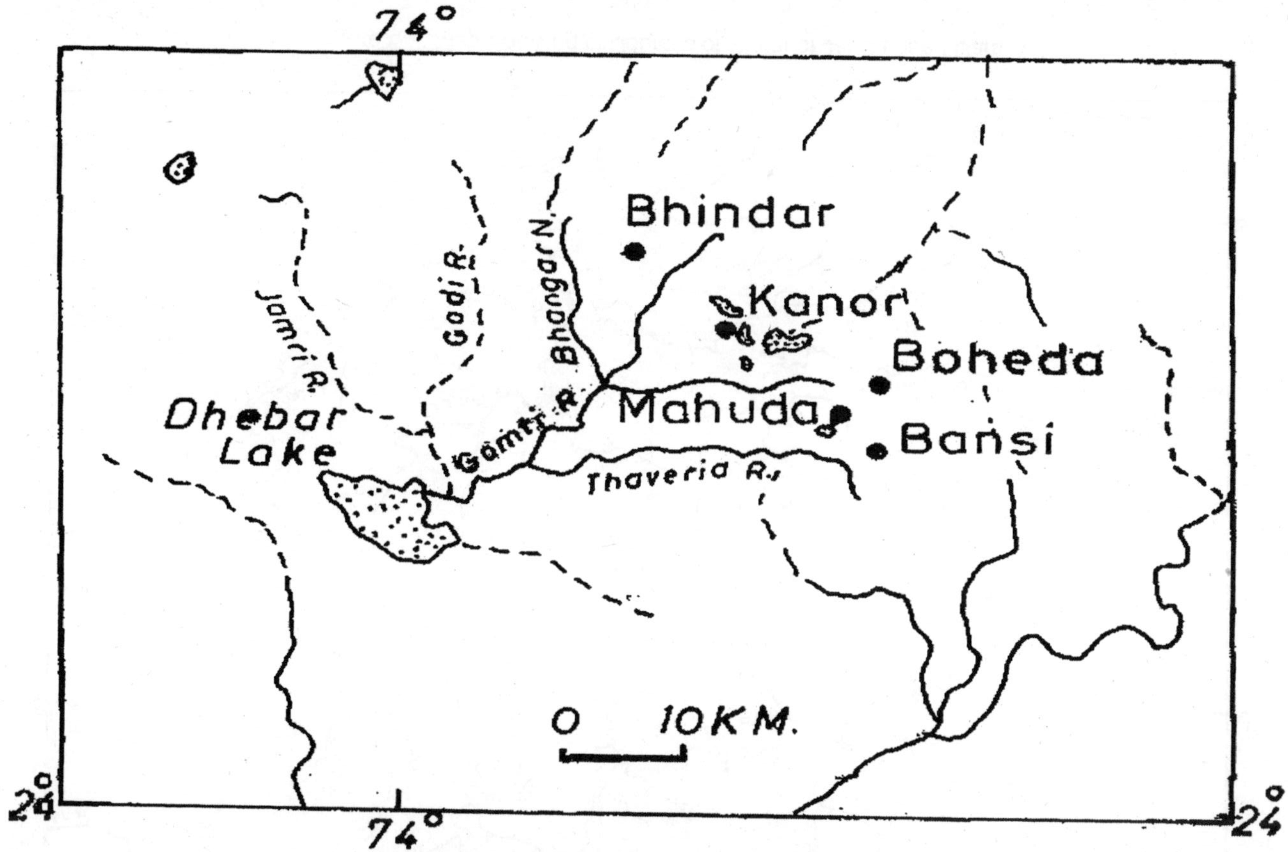

Figure 6.3: Jai Samand Feeder Tanks

after an earlier manuscript.[10] It seems to have been damaged during the invasions of the Mughals.[11] This *prashasti* has 17 cantos (*sarg*); the first 12 record the history of Mewar from its inception till the rule of Maharana Jai Singh. The last five cantos; viz., 13 to 17 furnish us information on this dam and allied subjects which are of immense value. The work records every detail about the structure from the selection of site to the digging of the foundation and its size on completion.

The site for the embankment was chosen in the rear of the village Kentora in the Chhappan area, a dense hilly tract. In all 23 villages were submerged during the course of its construction namely, Gatora, Patan, Gamri, Shukodia, Bhatwara, Nak, Dingla, Saimal, Adhwara, among others. The land covered by these villages and also the surrounding land was most suitable for the construction of the dam. The mountain, locally called Maniyal, was in two segments (Figure 6.4). In between the two was a deep gorge through which the river Gomti flows with mighty force. The engineers impounded the river at this point by erecting an embankment barrier. After getting clearance from the Sisodia chief, construction was initiated on *Baisakh Shukla* 10, VS 1743/AD 1686 and the filling of foundation work (*pad puran*) was taken on the *Asadh Shukla* 5. To make the substratum stronger, the land was dug to a depth of 6.7 m. and the water was drained away. The water was drained out with wooden pots (made of bamboo). Thus, on the solid bottom a foundation wall (*pad bandhan*) was raised after clearing the water from the subterranean level (Figure 6.5).

SIZE OF THE DYKE

The *prashasti* records the detailed measurements of each segment of the structure. It is 30.5 m. in length and the height measures 31 m. Its breadth at the base is 366 m. The length of different sections of the dam is recorded, which comes to 608 m. It has a uniform breadth of 30.5 m. at all points. It appears that the figures for length are misleading and there is a strong possibility that the

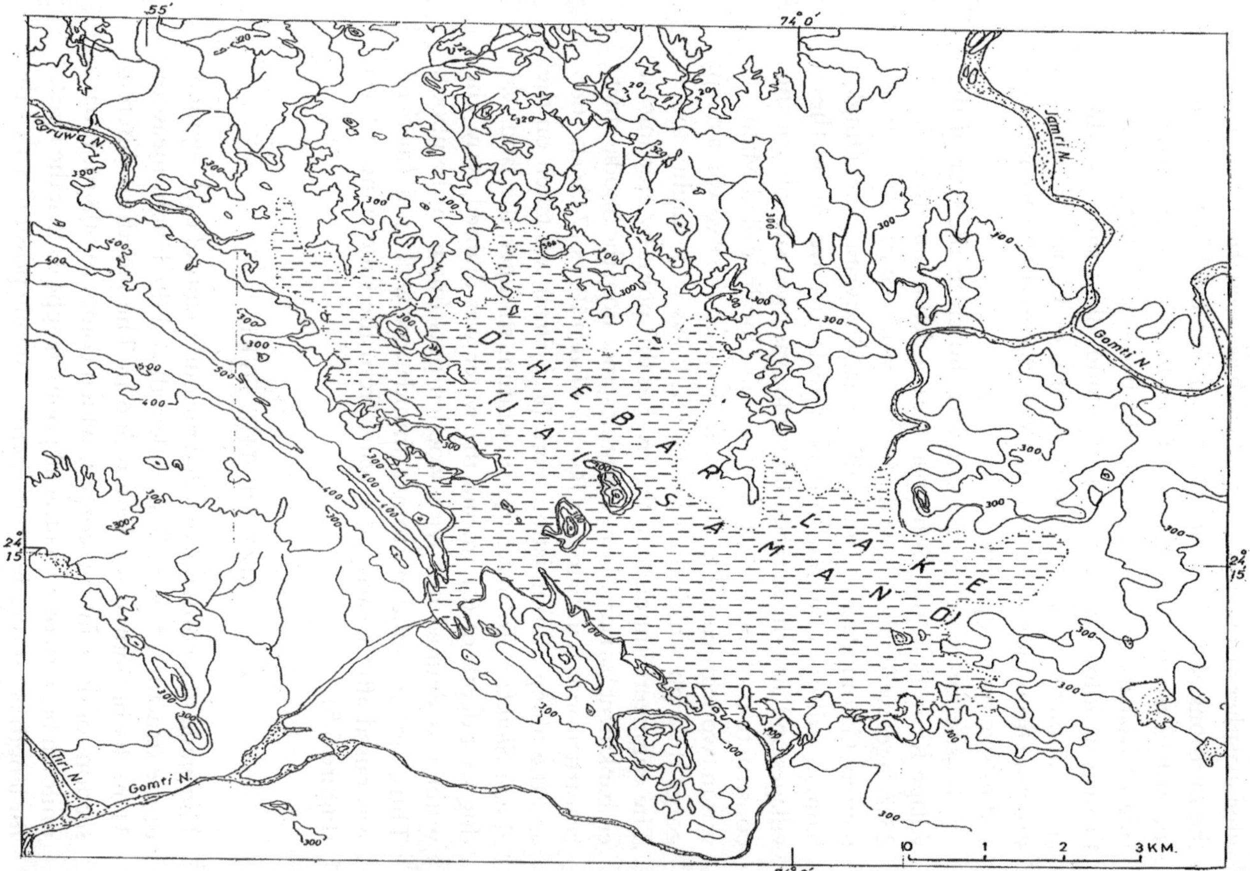

Figure 6.4: Jai Samand Dam and its Environs

Figure 6.5: Front view of the Jai Samand Dam

figures are read erroneously by a modern writer.[12] The correct figures can be known only after the scrutiny of the original manuscript.

It is said that the completion (*pratishtha*) ceremony was performed in the month of *Asadh* (June–July) of VS 1744/AD 1687 but work on the dam continued for a long period. The *prashasti* tells that the circumference (*pradakshina*) of the dyke was 128 km. at the time of its completion but it reached 400 km. at the time of the demise of Jai Singh in VS 1756/AD 1699. This shows that it had surpassed the Raj Samand Dam in size.

Modern authorities, too, have recorded the measurement of this structure in feet. In addition we have obtained detailed measurements by undertaking physical surveys. All the figures are grouped together in the following table, converting all readings into an identical unit, i.e. metre, for the purpose of comparison:

Measurement	Erskine's Gazetter	*Vir Vinod*	Ojha	Obtained by measurement of the structure (in mts.)
Length	381.86	382.47	366.61	371
Height	35.38	33.05	–	–
Breadth (base)	21.35	–	21.35	–
Breadth (top)	4.88	32.02	35.38	37.5
Height of back-wall	30.5	38.71	–	–
Depth	–	–	31.11	42
Depth rear	–	–	–	101
Distance between two walls	61 m.	–	–	99.50
Length of second wall	283.34	–	396.5	–
Breadth of back wall at the base	10.67	–	–	–
Breadth at base top	3.66	–	–	–

The measurements recorded by modern authorities are broadly closer to our actual measurements for some sections but divergent for others. The length of the embankment wall broadly corresponds to the figure obtained by the actual measurement of the structure. In case of breadth at the top, the scale recorded by Erskine and Ojha do not match with those of Shyamal Das. Actually, the difference is quite wide. The depth recorded by Ojha is 31 m. while our measurement comes to 42 m. Overall, there is a broad agreement in figures with some exceptions. Some dimensions were difficult to check because of water in the dam.

The thickness of the dyke is the most important element in the whole structure as it is the part which faces most the force of water. So the engineers attached more importance to this section. To safeguard the dam from the force of water it was considered necessary to keep the line of thrust within the middle third of the cross-section of the structure. This principle appears to have been followed in this body (Figure 6.6).

To give additional strength to the main embankment wall, stairs were constructed projecting towards the tank (Figure 6.7). Atop the dyke floor are six cenotaphs (*chhatris*). In line of the *chhatris* are erected three more *chaukis* (sitting place), at a lower level. Thus, there are a total of 24 structures on the entire length of the dam and they are identical in scale. The size of each platform is 9.60 × 4.50 m. So, there are four stages in one row from top towards the tank. Nine stairs of 0.350 m. were built between the two, thus the total stairs come to thirty (Figure 6.7). The total projection between the wall and the tank was 12.6 m. which actually increased the span of the dyke. By adding it to the width of the embankment wall it comes to about 50 m.

The erection of a temple at the centre of the embankment wall, besides giving it an aesthetic touch was probably meant for providing additional strength to the barrage. On each side of the temple were raised three cenotaphs of identical size at equal distance which gives it a perfectly geometrical look. As noted earlier, in alignment with the *chhatris* were erected three resting places (*chaukis*) facing the tank. The last six sitting places (*chaukis*), at the corner of the tank, were decorated with elephants carved of white marble (Figure 6.8).

Figure 6.6: View of Stairs

Figure 6.7: Structures on the Embankment

Figure 6.8: View of Jai Samand Dam

A contemporary source mentions the expanse of the dam. It records its circumference as 140 km.[13] According to an estimate of early twentieth century, 'its length from north to south is about 14.4 km. and its length varies from 1.6 to 8 km. It receives rain water from catchment area covering 1,766 sq. km. The area of the lake is 53.76 sq. km.[14] but a contemporary document gives 225.28 sq. km. which appears correct.[15] There is a strong possibility of shrinkage in area in later times (Figure 6.9).

Cost of the Barrage

The *prashasti* offers us the consolidated amount spent on the construction. According to it the total cost came to Rs.18 lakh[16] but no details are offered. It is interesting that this cost was less than that of the Raj Samand built by Maharana Raj Singh.

Outlet of the Weir

The *prashasti* mentions that the engineers cut out an outlet (*oto*) near the village Kherad[17] which was big like a river (Figure 6.10). The author writes allegorically that the Sisodia chief brought water on the earth by erecting an outlet as the Bhagirath brought the Ganga on the earth. In fact, the flow of water was like a river. It must have irrigated a vast tract of land. But another contemporary document notices three outlets cut out in the hills of the east. These were primarily for the safety of the hydraulic structure in case the volume of water was beyond the capacity of the tank. Secondly, the distribution of water for the purpose of irrigation must have dominated the minds of the civil engineers and the existence of canals in the western direction of the weir confirms this. The villages located on both sides were irrigated. According to one estimate about 48.64 km. of cultivable area was irrigated by the water from this dam.[18]

Masons' Signature

Generally, artisans' marks are available in abundance in the medieval buildings of Mewar. We have come across different kinds of signs on various water structures too. These were engraved by the

Figure 6.9: View of Expanse and Extent of Lake

Figure 6.10: View of Outlet of Jai Samand Dam

artisans on the main body or on any part of the structure for two reasons: first, to identify with a specific guild or to differentiate from other groups. Secondly, most of the artisans were unlettered. But one of the characteristics of the Jai Samand dyke lies in the signature put by the artisans. Every stone slab on the stairs has the name of a mason. Whether each slab was signed by the stone-cutter himself or his name was engraved by a literate mason is not clear. Another pertinent question is what necessitated the civil engineers to allow the artisans to put their signatures on the slabs? (Figure 6.11) (A few names have been given in the Appendix, p. 143.)

An analysis of the names of the artisans demonstrates the composite and cosmopolitan character of the group. Both Hindus and Muslims worked together in the stone-cutters' workshops. Among the Hindus, people from different caste groups appear to have adopted the profession of a stone-cutter. People from different regions such as Bundi and those from the east, i.e. Dholpur, Mathura and other areas in the *subas* of Agra and Delhi, gave it a cosmopolitan character. This indicates caste mobility. Two conjectures may be made regarding the multiplicity of the names: first, the names on the slabs were of those stone-cutters who worked on the dam; second, these were the rural folk from the surrounding settlements who brought stones on the invitation of the Maharana, thus cooperating in the Maharana's endeavour. Therefore, their names appeared in recognition. The latter probability appears more plausible because every signed stone is not of equal size and has a rough surface.

Architects of the Hydraulic Structure

The rulers could only express their will to build dams, step wells or tanks and bridges across rivers and raise the money to spend on their construction; it was the architect or engineer who could translate their dreams into reality. The architects and engineers prepared a layout and ground plan of the structure. They were solely responsible for the entire structure, from the selection of

Figure 6.11: Signature of the Masons

the site to the completion of the building. The architects of the Jai Samand were Gajdhar Nathu and Jagannath.[19] They were engaged in other works too, which shows their expertise in the construction of water monuments. Moreover, the stability of this dam over the last three centuries is proof enough of the skill and competence of the engineers in hydraulic engineering.

Managers of the Construction Work

The *prashasti* records that millions of diggers, artisans and masons were engaged in the construction of the dyke but it does not give the precise number of the workforce. The speed with which the task was completed itself indicates employment of a large number of labour. The dam was ready for operation in a record time of 13 months while more than 7 years were spent on the Raj Samand Lake; this was a remarkable feat as the former was more than double the dimensions of the latter. The officials who were supervising the work were Ranawat Bakhat Singh, Govardhan, Pancholi Anandidas and many others.[20]

The Jai Samand constructed by Maharana Jai Singh was really an enormous waterbody both in terms of its extent and size. It receives the water of four rivers but according to local tradition waters of nine rivers and ninety channels contributed to it. The size, both in the height and breadth of the embankment, was decided taking into account the volume of water and the force which was to be faced by the embankment wall. Secondly, it cost the state treasury Rs. 18 lakh, which was eight times less than that of the Raj Samand which was built at a cost of Rs.14,472,232. This encourages us to draw the conclusion that Jai Singh fulfilled the challenge thrown to him by his illustrious father Raj Singh, of constructing a bigger dam and that too at a much lesser cost.

NOTES

1. Jai Singh was born on 15 December 1653 (*Vir Vinod*, vol. II, pt. I, p. 645) and was about 23 years at the time of completion ceremony of the Raj Samand.
2. The village Dewali was part of *pargana* Girwa and its estimated income was Rs. 3,001 during the time of Maharana Raj Singh (*Maharana Raj Singh Pargana Bahi*, op. cit., p. 1). Shyamal Das records its length indicating two corners of the site, cf. *Vir Vinod*, op. cit., II, p. 667.
3. Ojha, op. cit., II, p. 903. The destruction of the catchment in the past few decades has increased the rate of siltation. It threatens the very existence of the dyke. There was a silt trap between the Madar tank at Thur ki Pal village but has now been destroyed. At present the silt flows into the lake (*The Dying Wisdom*, op. cit., p. 164).
4. The village Thur was a part of *pargana* Girwa and its estimated revenue was Rs. 800. It was located in the mountainous area (cf. *Maharana Raj Singh Pargana Bahi*, op. cit., p. 2).
5. *Vir Vinod*, II, pt. I, p. 645; Ojha, op. cit., p. 103.
6. Shri Khetal, 'Udaipur Ki Ghazal', in *Shrinagar Varnatmak Hindi Padya Sangrah*, ed. Muni Kanti Sagar, Surat, 1948, pp. 13–21.
7. Nainsi, *Khyat*, I, pp. 33–4.
8. G.H. Ojha, *Udaipur Rajya Ka Itihas*, II, Jodhpur: Rajasthani Granthagar, rpt., 1999, pp. 593–4.
9. Ibid. An anecdote current among the local people was told to the present author. According to it, Dhebar, probably a *Bhil*, was a village chieftain and the barrage was constructed with his co-operation. He was assigned the task of the supervision of the dyke. After its completion, the chief enquired from the Maharana about the name of the dam. The Sisodia chief replied that it would be called by two names, viz., Dhebar and Jai Samand. But according to a *ghazal* composed in VS 1757/AD 1700, Dhebar was the name of a village. Cf. Khetal, Udaipur ki Ghazal, op. cit., p. 20.

   After the discovery of the *Jai Singh Nrapa Prashasti*, an epic composed primarily on the Jai Samand Dam, the controversy over the name Dhebar has been set at rest forever. It was the name of a village and the place where the embankment was erected was known as the *Dhebar ka Naka*. The name of the village was changed to Jai Nagar (cf. Shakti Kumar Sharma, 'Shakunt', 'Jai Singh Prashasti men Jai Samudra-Nirman ka Vivran', in *Rajasthan ka Aitihasik Sanskrit Sahitya*, ed. K.L. Sharma, Jodhpur: RORI, 1996, pp. 70–5.

10. The poet writes in the colophon of the work that Maharana Amar Singh II had inspired him to recompose the *prashasti*. This clearly means that the earlier version composed during the time of Maharana Jai Singh could not get much publicity, probably because of the passing away of the Sisodia chief. In the meantime, the manuscript was probably destroyed. So later, at the wish of Maharana Amar Singh II (1698–1710), the *prashasti* was recomposed by the poet. But this epic could not be inscribed on stone slabs like the *Rajprashasti*. This remained in manuscript form and could not be traced before the 1960s. The academic world is indebted to Prof. Mool Chand Pathak for having discovered this rare manuscript. A copy of it is still preserved in the Asiatic Society, Kolkata. See, *Rajasthan ka Aitihasik Sanskrit Sahitya*, op. cit., pp. 70–5. But the text has not been edited and published so far. When Ojha was writing his book on Udaipur he had seen a copy of the *prashasti* but could not consult it. He found it similar to the *Rajprashasti* (cf. op. cit., p. 904).
11. Shakti Kumar, op. cit., pp. 70–5.
12. Ibid., since the length obtained by actual measurement comes to 371 m. Therefore the length recorded by Shakti Kumar in his article is misleading.
13. *Phutkar Khyat*, Kaviraja Collection, Granth No. 109, f.334(a), Netnagar Shodh Sansthan, Sitamau.
14. Erskine, op. cit., p. 8.
15. *Phutkar Khyat*, op. cit., f.334(a).
16. Shakti Kumar, op. cit., p. 74.
17. Ibid.
18. Erskine, op. cit., p. 9.
19. *Rajasthan ka Aitihasik Sanskrit Sahitya*, p. 73.
20. Ibid.

APPENDIX

Names of some of the Masons

1. Jago of Bundi
2. Duda of Bundi
3. Ratno of Bundi
4. Tilo of Bundi
5. Rupo Purbo
6. Palad Purbo
7. Lalo (*Khalisa*)
8. Senal Sompura (*Khalisa*)
9. Pom Chatar (*Khalisa*)
10. Rinchhod Sompura
11. Kanisa Mapra
12. Lalo Bhati
13. Ganes Somchand
14. Jogi
15. Jagu Siray
16. Suharji
17. Geha
18. Udo
19. Ranchhod (of village) Thaysado
20. Sero Fakiro (*Khalisa*)
21. Karim Gauri
22. Seru Piru
23. Piroj Gauri

PART II

UNDERGROUND WATER

CHAPTER 7

Irrigation in Mewar: Wells

Mewar depends for its water on the monsoon as well as on the sub-soil or underground water. In case of good rainfall the region is in a better position than other parts of Rajasthan in terms of water availability. It gets fairly regular rainfall from the Indian Ocean in the south-west and also from the Bay of Bengal in the south-east. If the south-western monsoon from the former direction fails early, then the winds from the south-east come to the rescue later in the season.[1] The average rainfall varies in the different areas of the state. The average annual rainfall in the central part, including the capital, since 1880 has been about 622 mm. The maximum fall recorded in any one year was nearly 1,130 mm. in 1893, and the minimum just under 254 mm. in 1899.

The rainfall in the south-west is usually more than in the capital. Thus the averages for Kherwara and Kotra come to 673 and 800 mm. respectively. The gap in the average rainfall is sometimes quite wide. While the former received a maximum of 165 mm. in 1899 the latter received a maximum of 1,549 mm. in 1893. Kumbhalgarh receives more rain than Kotra. The north and north-east of the state receive slightly less than the region around the capital.[2] This shows that the distribution of rainfall in Mewar is uneven. Systematic rainfall records which date back to the mid-nineteenth century do not suggest any significant change over the past century and a half.[3] It encourages us to presume the same average of rainfall for early centuries too. This presumption gets support from evidence furnished by Nainsi, a celebrated statistician of seventeenth-century Marwar who has compiled an exceptionally rich data even for Mewar. He records that the mountains such as the Machhawala and Jarga located in the south-west contain enormous water reserves and dense forests.[4] This is the zone of fertile tract where wheat, rice, gram and *urd* are produced in abundance and mango trees and wild plants are in plenty.[5] This shows that the south-west attracted higher rainfall than other parts of the state in the seventeenth and in earlier centuries too. Thus we may infer that the water-table in the south-west was higher than the north-east and the central parts of the state.

Since earliest times wells were excavated to tap underground water for which we have evidence for Mewar. Gradually wells became an important means to exploit sub-soil water for irrigation. The digging of wells in India was prevalent both for domestic requirements as well as for irrigation purposes. It was the simplest method to extract underground water by the agriculturists and the urban dwellers. Rajasthan was not an exception in this sphere. People of the desert (in the north-west) and in the non-desert or fertile tracts in the south-east excavated wells to meet their requirements of water. To some extent it is true that the wells dug out in the desert areas could only fulfil household requirements as sufficient sub-soil water is a prerequisite irrigating small plots or large tracts. The depth of the underground water was therefore of primary importance. This was the determining factor which would decide the feasibility of a well for the purpose of irrigating an adjoining agricultural field.

The underground water-table depends on the amount of rainfall and the presence of surface water (river and *nallahs*). Besides, rivers and *nallahs*, tanks also contribute in raising the sub-soil water level near the surface. Thus percolation of water from the above means and the sponging of the rainfall enlarged the underground water reservoirs.

Well irrigation was widely prevalent in Rajasthan[6] and elsewhere in India.[7] In many parts of Mewar, such as the north and west, there was only a thin layer of soil covering the rocky sub-stratum. The wells reaching down to the rock could not give year-round supply of water. It was necessary to dig through the rock to reach the underground springs in these areas. This makes it laboursome, enhancing the cost of the wells. These wells remain serviceable for a short period and then dry up. Only very small pieces of land can be irrigated by these wells. In case of failure of rains or insufficient showers, these wells become almost redundant or provide insufficient water for irrigation. This scenario compels us to infer that rivers irrigated a major part of the cultivable land in Mewar.

Rivers and channels provide a baseline for the digging of wells in the state. The river courses all along are best suited for this work. The water spreads all around from both the banks of the rivers and raises the groundwater level nearer to the surface. This is called *seja* in local parlance. Most of the wells were dug in these locations particularly on both sides of the rivers, since these were quite inexpensive (Figure 7.1). In fact, these are shallow wells and the undercurrent is tapped quite near the surface. This characteristic gave the wells the name *seja kuwa* (shallow well).[8] The availability of water depends on the fast-running undercurrent and its depth varies from 7.6 to 9 m.

Another kind of well called the *akhara* was quite deep but it was expensive to dig and gave much less water than the *seja* wells. The depth of the *akhara* wells range between 13.7 and 15 m.[9] Those kinds of wells are shown in following map (Figure 7.2).

A third group of wells which have escaped the notice of historians and gazetteers is found around the tanks. They also constitute an appreciable share in the aggregate of wells. In our physical survey we found about 10 wells around a tank which irrigated a large tract (Figure 7.3). As the wells excavated on the banks of rivers received percolated water, similarly that water level of these wells too depended on the tank water.

The inscriptions of early medieval times from various parts of Rajasthan employ three different terms for wells: *dhimda* or *dhivda*, *vapi* (step well) and *araghatta*, *araghata* or *arahata*.[10] Excluding the *vapi*, the difference between these appears to be related to the operational variations in well irrigation. The distinction between the *dhimda* and an *araghatta* becomes explicit in an epigraph. The difference between the two is related to the amount of land irrigated by them. The *dhimda*[11] in a well without a Persian-wheel

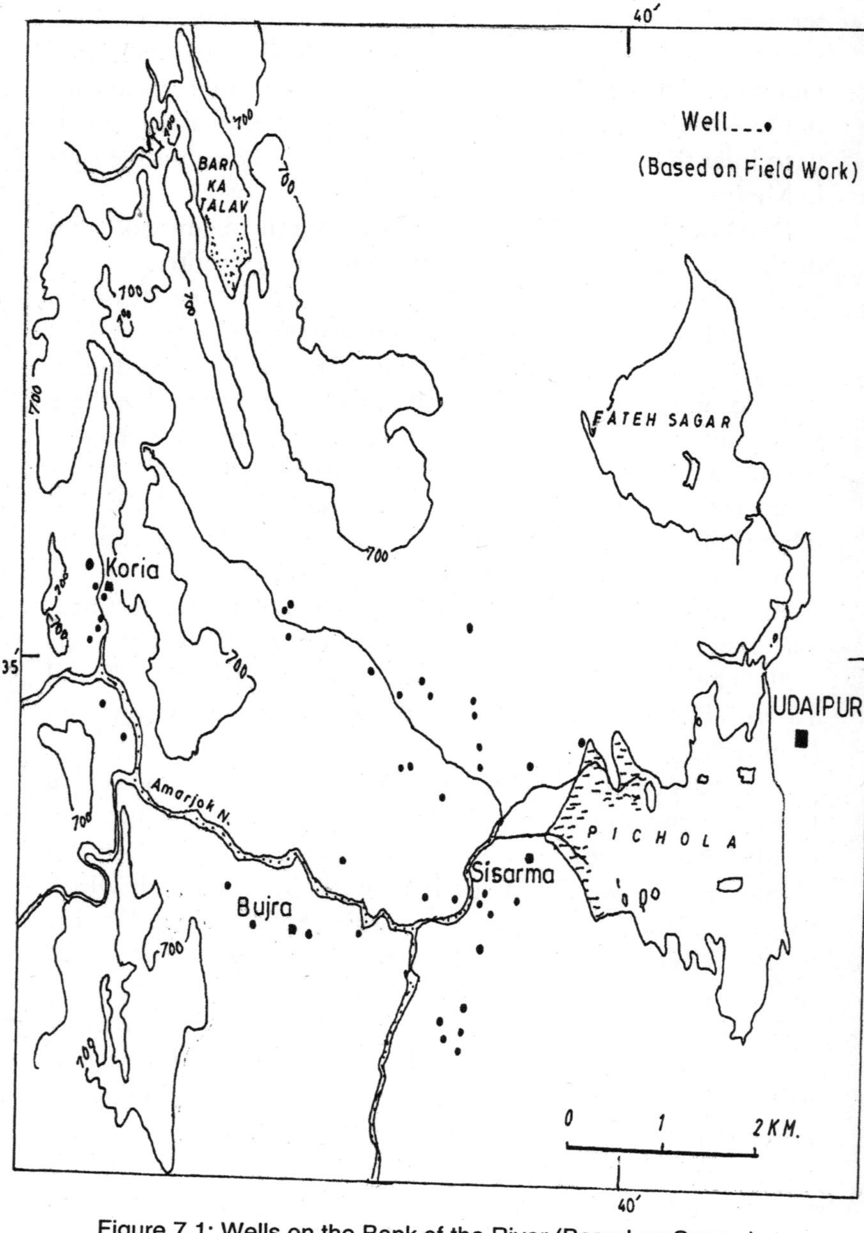

Figure 7.1: Wells on the Bank of the River (Based on Survey)

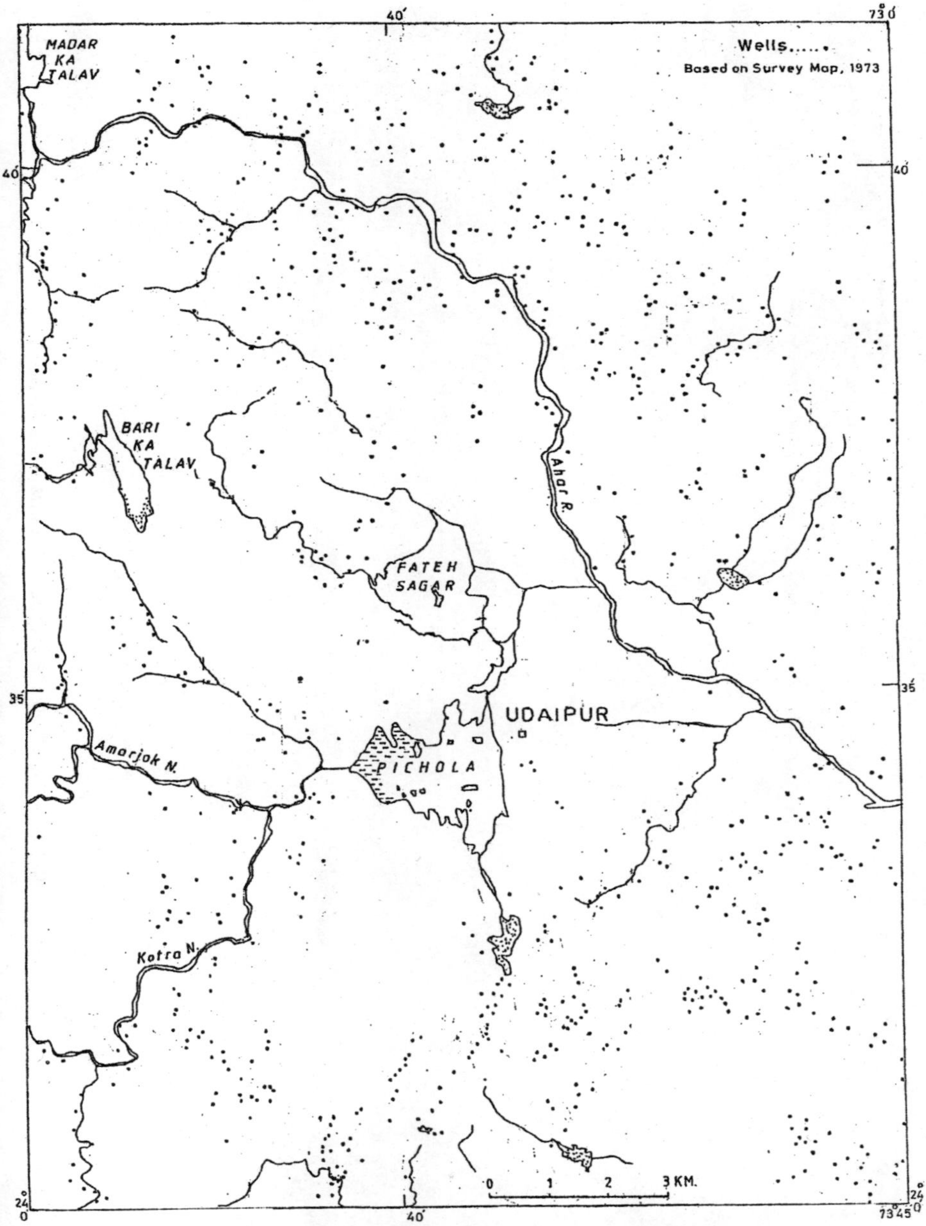

Figure 7.2: Map of Wells in and Around Udaipur

Figure 7.3: Well on the Corner of *Kamalwala Talab*

which irrigates less amount of land than the *araghatta* operated well. Obviously then the levy on the produce of land irrigated by the former would be less than from the latter. From these the relative importance of *araghatta* over the other apparatus becomes obvious.

B.D. Chattopadhyaya has compiled a list of different kinds of wells which irrigated different regions. According to him, two villages, namely Virpura in Chhapan and Ahad of Udaipur, had wells with *araghatta*. Another village Ghagsa was watered by a step well.[12] It becomes clear from the reference about the *araghatta*[13] that its operation was distinct from that of an ordinary well. It also establishes the existence of two kinds of wells in Mewar. Broadly, two kinds of water-lifting apparatuses on wells were employed: *charas*[14] and the Persian-wheel. The reading of the contemporary revenue records of Mewar demonstrates two clear-cut blocks for these devices. The *charas* was used to lift water in the *parganas* of the east and north while the Persian-wheel was operated on the wells dug in the southern and the western part of the state. It is also pertinent to note that in the *pargana* of the north-east and the central part of Mewar more than one *charas* was employed on the wells. This apparatus was used as a rule on the deeper wells. In the *parganas* of south two *charas* were generally employed on one well. The Persian-wheel was widely operated in the entire territory of Mewar. This means that the water was quite close to the surface.

This is a very interesting phenomenon which implies that the extensive tracts of fertile land in Mewar were found in the areas of deep alluvial soils which were created on both the sides of the rivers through the percolation of their water. In most such areas, water could be tapped at no great depth by digging wells.

SELECTION OF SITES FOR WELLS

The selection of site for digging a well assumes significance because of the involvement of high costs in hilly terrains. For this, the accumulated knowledge of villagers played a very important role.

They knew about the topography of the area and could guess the availability of underground water. This practical knowledge was transmitted from one generation to another. Besides the local people, there were water-diviners in every area who had expertise in identifying spots suitable for digging wells. They were known by many names, such as *panikals* in Gujarat and *pathrot* and *takari* in Rajasthan. They were skilled in telling at what depth a spring would be detected. They chose sites with care and skill and their judgement was generally accurate. Their services were obtained to escape unnecessary expenses of failing to detect a spring.

MATERIAL FOR THE CONSTRUCTION OF WELLS

A physical survey of some of the areas of Mewar reveals that the majority of the wells are of masonry and that unlined or *kachcha* wells are extremely scarce. The obvious reason is the availability of stone-rubble in great quantity, especially in the hilly areas. Other kinds of wells are found only in non-hilly areas where stones are not available. Besides this, these unlined wells were mainly dug out in the rainy season near the river beds or seasonal *nallahs* where the water level reached near the land surface.

The masonry wells were built with stone slabs rather than bricks. No brick-well could be found in the entire territory of Mewar. Stone slabs or rubble could easily be found in the surrounding hilly areas, and in non-hilly open areas stones were brought from the adjoining or even far-flung *parganas* by bullock-carts. This would have certainly raised the cost of the well. The deep wells, *akhara*, appear to have been generally excavated in the hilly tracts where springs were detected at great depths. In these tracts wells were dug through the rocks to tap deep underground.

Neither any major construction work nor use of much lime-mortar was required in some of the wells. In these cases, the diggers hewed out the rock with chisel. In other cases, local techniques were employed to split the rock. One such well is found behind the Udai Sagar Dam in the vicinity of Udaipur.

This well seems to have been excavated in an early period[15] (Figure 7.4). It was dug out by cutting the rock down to the underground spring. This shows that this technique was employed by the villagers or diggers since an early period. A deep well known as *kalalia* was excavated in the middle of the city of Udaipur. A *ghazal* of the eighteenth century mentions that this well existed near the main market (*mandi*). It also mentions that it is as deep as the Pichhola Lake and is full of water like the dam (*pal*) of Bari, i.e. Jana Sagar dyke.[16] It shows the high water table in the very heart of the city which was the result of the existence of the Pichhola and Rang Sagar tanks in the vicinity.

AGENCY OF CONSTRUCITON OF WELLS

A large number of extant well structures spread over the entire territory of Mewar show that these were constructed during different periods of history. The location of a well obviously determines the purpose of its construction. Ordinarily, wells are found either near the agricultural fields or in the vicinity of populated areas, temples and other kinds of building complexes. Sometimes, they were constructed in deserted areas or along the routes. The wells found near or in the middle of the fields were certainly meant for irrigation. Others were for the purpose of providing water for drinking and other mundane uses.

The agencies involved in the construction of wells included individuals or peasants, group of peasants or persons, village communities, caste panchayats and the state. The category of well—unlined or masonry—was also a determining factor for the engagement of an agency. Shallow, unlined wells could be dug by an individual peasant or members of a joint family. The construction of masonry or rock-cut wells involved extremely hard work which could not be carried out individually. The peasants had to hire the services of skilled labourers or diggers who had expertise in the field of well construction. They were known as Kinia, Meghwal, Bhid and Ods in Mewar and also elsewhere in Rajasthan. Kinia is not a separate caste but people who had acquired expertise

Figure 7.4: Photo of Ancient Well behind Uadai Sagar

in digging of a well, irrespective of the caste they belonged to. But generally labourers from the families of Meghwal, Od and Bhid quickly learnt the technique and skill in this field.[17]

Innumerable extant shallow well structures can be seen along river beds and channels in different localities. Generally these are found in the fields owned by individual peasants. We do not have any means to know the identity of their diggers but circumstantial evidence shows that these were constructed by the peasant on whose land the well existed. Two possibilities about the mode of construction of well seem plausible. One, either the well was dug by an individual peasant himself using his family labour or the owners of land hired the services of professional diggers. They paid the professionals either from their own pocket or borrowed money for the purpose.

The evidence of *pargana* Mandalgarh is of extreme significance which enlists the names of the villages along with the number of houses, ploughs and wells. Four villages, namely, Adwari, Namatpuro, Ranpuro and Ulgam, had 59 houses of agriculturists who owned 39 ploughs and 45 wells (*kuwa*).[18] In one village two households cultivated their fields with the help of two ploughs and irrigated their fields from two wells. The cultivators of the village Adwari, who were 11 in number, irrigated their fields from 15 wells. Each cultivator owned more than a well. Overall, 59 households used 45 wells for watering their fields which clearly shows a favourable cultivator-well ratio in the *pargana*. The ratio between the cultivator and well came to 1:0.76 in 1833. This makes us infer that ordinarily an individual peasant could own his own irrigating monument.

We have come across another kind of well in Mewar which belonged to a group of peasants whose fields were adjoining each other. This type of well is located on the outskirt of Udaipur on the Jaipur–Udaipur highway. There was an arrangement among four or five peasants to share the well water for watering their fields. This well had Persian-wheel fixed permanently on it. All the peasants had their own separate chain of earthen pots (*mal* or *mala*) and a pair of bullocks to draw the water from the well to irrigate their fields. The well and the wooden machine were for common use. In this arrangement the well seems to have been constructed jointly, the expenditure incurred on it probably borne by all the concerned parties. We do not have any evidence indicating involvement of a *sahukar* as in the case of Gujarat.[19] Mewar too had a large number of wells constructed with the help of moneylenders. In the Kachhwah territory, a large number of wells were constructed with the assistance of the Bohras. This was a marked feature of the agrarian policy of the state, because it was always extremely difficult for the peasants to pay the required amount from their own pocket for this purpose.

The state was also involved in the construction of wells. This gets corroboration from the evidence contained in the *Bilara Sangrah*. Thus, Maharana Amar Singh granted agricultural field along with a *dhimra* well to Chaudhary Chanda in VS 1674/AD 1617.[20] Similarly, a *rahant* was granted to Chaudhary Rohitas in VS 1660/AD 1603.[21] Besides these, innumerable documents show that wells were built in rural as well as in urban areas. State-ownership over the wells is well established because the ownership was sometimes transferred to certain individuals by the ruler. It can be inferred that generally these wells were dug in the *khalisa* areas. All the expenditure on wells with water-lifting apparatuses was paid from the state exchequer. The rulers sometimes are seen constructing wells in *jagir* and *sasan* (revenue-grant villages) areas for the benefit of the peasants. The interest of the state in the construction of wells is visible in the instructions given to the *thikanedars* and the officials through the *parwanas*. A large number of such letters testify to the encouragement by the rulers to individual peasants to dig well in their respective villages and in return offer concessions in the revenue demand.[22]

The *thikanedars* were owners of a large number of villages enjoying hereditary rights over them. Hence, they took keen interest in the construction of water monuments in their territory. They had large areas under their control and to improve the agriculture in their *jagir*, they built wells in large numbers. The landlords of the *thikana* Bedla and their wives constructed wells, step wells, *kunds* and tanks.[23] Similarly, an enormous tank was excavated in Kanor by the *thikanedar*. On the corner of it, six wells were constructed to tap the sub-soil water. This entire water complex was used for watering the surrounding fields.[24] This is sufficient to show that a large number of wells were built by the *jagirdars* of Mewar in their respective estates. We do not have any means to know whether they themselves financed the construction of such wells or borrowed from the *sahukars*. But one thing is certain—which is corroborated by the documents—that the *thikanedars* always remained interested in the extension of agriculture in their estates. They always encouraged castes known to be skilled in cultivation to settle in the villages of their *thikana*. The Chaudharies of Bilara belonged to the Sirvi caste[25] and were expert cultivators. Maharana Amar Singh invited Chaudhary Rohitas and Bhagwan Das and granted them wells to cultivate the land of village Kaylana. This also shows that the ownership of well vested in the state.

The physical survey conducted by the present author shows the existence of some wells named after particular castes. In village Bari, located behind the *Bari ka Talab* or Jana Sagar, there is a step well, known as the *Rayno ki Baori*. Rayno is probably a lower caste but according to local populace it means a kind of fruit which grows there. It is reported by local people that it was formerly known as the Maharana well. It was constructed prior to that of tank of *Bari*. Near this step well is a regular well which is still in operation. *Rahat* was used to lift water from it. It is reported that it irrigates more than 20 *bighas*.[26] Both these structures belonged to the village population consisting mainly of the Bhils.

There is another structure in the vicinity of the above wells, known as *Kunjaro ki Baori*. The Kunjar[27] were lower caste people who dealt in miscellaneous merchandise. The Persian-wheel was employed for watering the fields. A canal from the Jana Sagar runs parallel to this *baori*. The water from the dam percolates and raises the water level of the locality.

The above extant structures show that these were not constructed by an individual but by caste panchayats. This means the involvement of this social institution in the construction of wells for particular caste members.

The state took keen interest in irrigation because irrigated lands provided higher land revenue. The rate of revenue demand here was always higher than in case of the dry land. The revenue schedule records separate revenue rates for *piwal*, i.e. irrigated or watered lands and ordinary lands. In *pargana* Mandalgarh, the rate of cotton grown on *magra* (hilly tract) and *gorma* (irrigated land) was distinct. The rate on the latter land was one and half times higher than the former.[28] Besides it, the state also encouraged cultivation by irrigation. Maharana Sangram Singh granted 50 *bighas* of *parat* land to Chaudhary Bhagwan Das in village Kaylana of *pargana* Godwad. He was specifically instructed to dig a well (*kund*) and cultivate the field. This clearly indicates the state policy of promoting well-irrigation.[29]

The sources do not provide information about the number of wells during the medieval period but we get figures for wells during the colonial period. According to one estimation there were 25,000 wells in the settled districts and about 100,000 in the entire state of Mewar.[30] The state has an area of 32,489 sq. km. which means there were 7.88 wells per sq. mile. The perceptive comment of Tod also supports the contention that the number of wells in Mewar was much higher than in the southern districts of Marwar as the water is near to the surface.[31] The paucity of data for the pre-colonial period puts restrictions on working out a similar percentage for the period but one may say with due caution that well irrigation was more widely prevalent during the medieval period than the colonial period.

The irrigation devices used for lifting water depended on the water level beneath the surface. Where the sub-soil water was near the surface the *dhenkli* used for raising the water. This apparatus was employed presumably only on wells located on the sides of a river and which were styled as *seja* wells in Mewar.[32] In fact, it was a rope-bucket technique without any mechanical aid. But since this device was employed for watering small plots growing crops or vegetables,[33] there is a strong possibility that either small fields were being irrigated or some other device was operated on these wells. The *dhenkli* involved a man dipping his bucket into the water and emptying it into a small tank or trough from where the water went through *nalis* to the fields.

Another device was the *charas* which was employed in areas where the water-table was low. The author of *Vir Vinod* writes that this device was used in the north-east and central parts of the state.[34] Qaisar writes that the *charas* was an improved variety of the rope-bucket-pulley contraption in which manpower was replaced by a pair of oxen. Now it has assumed the status of a very specialized method meant to draw water for irrigation and is known as the *charas*. The *charas* was a huge bag made of leather which could draw immense water from the well in one round. Based on the analysis of Mughal paintings Qaiser draws the conclusion that this contraption was operated by two men, one used to stay on the well for emptying the bucket and the other positioned himself on the rope. The bullock track was like a ramp. The length of this track gives an idea of the well[35] which incidentally also confirms the statement of James Forbes who says that the length of the path corresponds with the depth of well.[36]

Another method of lifting water was the Persian-wheel. The author of *Vir Vinod* writes that this method was widely employed in the southern and western part of the state.[37] In the Mewari documents the terms employed for this apparatus are *rahat* and *arhat*. In Persian the term *saqiya* was used which Qaisar equates with *jal yantra* (water machine).[38] Here he highlights the employment of the wheel as the basic component of the apparatus which was operated through a gear system. The third important part of the apparatus was the use of animal power. Qaisar traces its evolution in his article.[39] The one thing that can be said with certainty is that the terms *arhat* and *rahat* do not represent the first and second stage of the *noria* as suggested by Qaisar. It appears that this term came to light after proper development of the Persian-wheel. The simple argument is that the terms *rahat* and *arhat* are still in operation, indicating the Persian-wheel and the *ghatiyantra* have gone out of use (Figure 7.5 A & B).

A

B

Figure 7.5: Photos of Wooden Persian-wheel

NOTES

1. K.D. Erskine, *Rajput Gazetteers: The Mewar Residency*, pp. 11–12.
2. Ibid., p. 12.
3. Erskine compiled rainfall data for the city of Udaipur, Kherwara and Kotra. As per his figures the city of Udaipur received 622 mm. of annual average rainfall during the last two decades of the nineteenth century while the areas located in the south-west, that is Kherwara and Kotra, received higher rainfall ranging between 673 and 800 mm., respectively. Op. cit., pp. 11–12.
4. Nainsi, *Khyat*, pp. 40–4; *Vir Vinod*, I, pp. 105-6; Ojha, op. cit., I, p. 2.
5. Nainsi, op. cit., pp. 40–4.
6. Well irrigation was widely in vogue in the desert state of Marwar. Its importance became more prominent during the failure of rains. A large number of wells were in operation in Marwar in the seventeenth century. The *parganas* located in the east and south-east like Merta, Sojhat and Jaitaran contained the largest number of wells. These *parganas* were located on the fringe of the Aravalli range. The average number of wells per sq mile worked out was between 1.44 and 1.46 which in 1929–30 was 1.78 wells per sq mile. This shows the importance of well irrigation in Marwar (cf. B.L. Bhadani, *Peasants, Artisans and Entrepreneurs: Economy of Marwar in the Seventeenth Century*, Jaipur: Rawat Publications, 1999, pp. 41–51). Erskine writing about Mewar says that the chief sources of irrigation were wells, of which there were said to be at least 25,000 in the settled districts and about 100,000 in the entire state. The character of the sub-soil, however, rendered the construction of wells expensive and labour intensive. A layer of hard rock lies within a few feet of the surface, and blasting alone enables the cultivator to get through this obstacle to the water beneath, and even then the real spring is rarely or never found. The well is filled by a more or less rapid system of percolation; the deepest and most expensive wells often run dry after being worked for a few hours, and the owner must wait until the supply is renewed. The extent of land irrigated by each well in a season averages about 5 *bighas* or more than 2½ acres, op. cit., I, pp. 41–7.
7. Irfan Habib writes that in the sixteenth and seventeenth centuries, the arid regions of Punjab were turned into rich agricultural areas

because of the construction of a large number of wells. Presidential Address: *PIHC*, Varanasi, 1969, pp. 153–4. Also see Chetan Singh, *Punjab in the Seventeenth Century*, Delhi: Oxford University Press, 1988, pp. 93–8.

8. *Vir Vinod*, I, pp. 118–19. In Marwar two kinds of wells were in operation: *chanch* and *kosita*. The former was a well with the *dhenkli* mechanism. The *dhenkli* based on the lever principle could only be used on shallow wells where the water level was not more than 1.83 to 2.75 m. below the surface. The *kosita* well most probably had the *charas* operated on it. Presumably *kosita* was not a shallow well in the strict sense, since two different kinds of apparatus, viz., *charas* and sometimes the Persian-wheel were used. Marwar historian Nainsi records the depth of the water of *kosita* wells with a Persian-wheel as 13.73 to 25.60 m., recorded in terms of *hath* and *puras*, and without a Persian-wheel as 4.58 to 13.73 m. (cf. B.L. Bhadani, op. cit., p. 42).
9. Erskine gives a description of the kinds of wells. He writes that:

   On either side of the rivers and streams, wells are numerous and least expensive. They are called *seja* or spring wells from the belief, founded on the abundant flow of water, that the spring is reached, but the constant supply seems to be solely due to more rapid percolation. *Akhara* is the name given to the other kind of wells which are avowedly percolation-wells; they are sunk much deeper and are therefore much more expensive and the supply of water is more precarious than in *seja* wells. But they are necessarily most prevalent, the others being confined almost entirely to the banks of rivers. The average cost of a *pakka* or masonry well varies from Rs. 640 to 800 while that of a *kachcha* or unlined well is about Rs. 400. Op. cit., pp. 46–7.
10. B.D. Chattopadhyaya, 'Irrigation in Medieval Rajasthan', *Journal of the Economic and Social History of the Orient*, vol. XVI, pts. II-III, Leiden, 1973, pp. 298–316.
11. Ibid, p. 305. In Marwar the wells of same nomenclature existed in the seventeenth century. The term *dhimra* or *dhibra* carries two meanings, that is, a well and Persian-wheel. The authority on Marwar, Nainsi specifically mentions the combined word as *arhat dhimra*, and sometimes splits it into two. It can be inferred that *dhimra* or *dhibra* was a kind of well with a Persian-wheel. The water-

level of *dhimra* wells mentioned in Nainsi as 12.8 to 25.6 m. or 7 to 14 *puras* confirms that only this apparatus could have been used effectively for drawing water (cf. Bhadani, op. cit., p. 45).

12. Chattopadhyaya, op. cit., pp. 307–8.
13. In the majority of the translations of early medieval Rajasthani inscriptions, the term *araghatta* has been translated as either 'mechanical well' or 'Persian-wheel'. According to Irfan Habib the *araghatta* represented pre-Persian-wheel technology and operated on the water surface (cf. 'Technology Changes and Society: 13th and 14th Centuries', op. cit., pp. 12–19). The evidence from early Rajasthan does not agree with it. On the strength of the available evidence Chattopadhyaya has cast doubts over the meaning of the *araghatta* as *noria* (cf. op. cit., p. 306).
14. In Gujarati, *charas* is called *kosa* (cf. H.H. Wilson, *A Glossary of Judicial and Revenue Terms*, Delhi: Munshiram Manoharlal, 1968, p. 295). In Rajasthani, the term *kos* appears to have derived from a unit of measurement, i.e. *kos*. It is taken to indicate the depth of a well. So the terms *charas* and *kos* were both employed for a leather bucket. In Marwar a well was identified by its water-lifting device, like the *kosita* well which gets its nomenclature because the *kos* (*charas*) was attached to it. For Mewar see observations of Erskine:

> Water is raised by means of the Persian-wheel (*rehant*), or when the spring-level is too far down for this contrivance, by the usual leathern bucket (*charas*) worked by a rope attached to a pair of oxen, and running over a wooden pulley. In shallow wells, the Persian-wheel is sometimes worked by the feet and is termed *pavti*. Other methods of lifting the water are by means of a *dhenkli* or an *indoni*. The former consists of a stout rod or pole, balanced on a vertical post and having a heavy weight at one end and a leathern bucket or earthen pot suspended by a rope to the other. The worker dips the bucket or pot into the water, and, aided by the counterpoise weight, empties it into a hole from which a channel conducts the water to the fields to be irrigated. The *indoni* is a basket covered with leather having a rope attached to each side; it is only used for shallow wells and reservoirs and is worked by two men, being merely dropped into the water and when full raised to the surface. Op. cit., I, pp. 47–8.

15. This well exists behind the Udai Sagar Dam near the canal. It is below the hill. Presumption can be made that it was used to irrigate the adjoining agricultural fields. An ancient temple also exists near it. I am grateful to my friend Dr. Jivan Singh Kharakwal of Udaipur for identifying the period of construction of the well.
16. Jaswant Sagar, 'Udaipur-Varnan Chhand', ed. Agar Chand Nahta in *Maru Bharati*, vol. 6, part 4, January 1959, Pilani: Birla Educational Trust, pp. 40–4.
17. Anupam Mishra, *Rajasthan ki Rajat Bundeyen*, op. cit., pp. 66–8.
18. 'Mewar Jagridaran-ri-Vigat', in *Majjhamika*, 16, p. 69. This document records the enumeration of houses, ploughs and wells for villages of *pargana* Mandalgarh in the following manner:

Name of the Village	House	Hal	Well (*Kuwa*)
Adwari	11	11	15
Namatpuro	7	6	5
Ranpuro	39	20	23
Ulgam	2	2	2
Total	59	39	45

19. David Hardiman, 'Well Irrigation in Gujarat: Systems of Use, Hierarchies of Control', in *Economic and Political Weekly*, vol. XXXIII, 20–6 June, Mumbai: Sameeksha Trust Publication, 1998, p. 1536.
20. Chaudhary Chanda was granted a *dhimra* well and a piece of land in village Kaylana of Mewar. This apparatus was of wood. This clearly demonstrates the Chaudhary was given a well to irrigate the agricultural field. This might have played a significant role in the extension of irrigated fields. (Cf. *Parwana* of Maharana Amar Singh, VS 1674, Bilara Collection; copies are preserved at the Natnagar Shodh Sansthan, Sitamau.)
21. The well with a Persian wheel was given to Chaudhary Rohitas by Maharana Amar Singh in VS 1660/AD 1603. The state ownership is implied in this case (cf. *Parwana* of Amar Singh of VS 1660, ibid.). A dispute between Chaudhary Bhagwan Das and Chaudhary Rohitas over the ownership of the *rahant*-well in village Kaylana was reported to the state. On this, the Prince Karan wrote to

Rohitas that the Maharana had instructed Bhagwan Das not to harass him and that he should be allowed to use the well without any hindrance. This *parwana* demonstrates the level of interest of the state in irrigated fields (cf. *Parwana* of Prince Karan Singh, son of Maharana Amar Singh, ibid.). Similarly, the Kachhwaha rulers were conscious of the importance of wells for increasing the land revenue income of the state. They provided loans to cultivators for the construction of wells. It was once reported to the ruler that 400 new wells in *pargana* Bahatri and another 15 in the Phagi were constructed.

22. In Mewari the term *kudo* is employed for referring to wells. In VS 1768/AD 1711 Maharana Sangram Singh granted *parat* land in village Kaylana to Chaudhary Bhagwan Das and instructed that he would cultivate it after digging a well there (cf. *parwana* of VS 1768, *Bilara Sangrah*, op. cit.). Similary, later Maharanas encouraged peasants to dig wells (cf. *parwanas* of various rulers compiled in *Mewar Ke Aitihasik Patte Parwane—Thikana Vijaypur Sangrah AD 1695–1907*, ed. Hukum Singh Bhati, Udaipur: Pratap Shodh Sansthan, 1993, pp. 113, 134 and 137).
23. 'Thikana Bedla ka Itihas', in *Majjhamika*, vol. 37, year 2002, pp. 4–7. This *thikana* is situated partly near the city of Udaipur but chiefly in the vicinity of Chittor and included 111 villages. The estate is held by the second senior noble of Mewar who is a Chauhan Rajput and bears the title of Rao. Bedla is situated about 6.4 km. north of Udaipur city on the left bank of the Ahar River. The other portion of the *thikana* is situated about 11.2 km. north of Chittor on the right bank of the Berach River, in the village of Nagari (cf. Erskine, op. cit., pp. 93–4). The *thakurs* of this *thikana* constructed numerous waterbodies in their villages both for irrigation and drinking.
24. These wells are found in our survey of Kanor and its surrounding areas conducted in the month of September 1999. This *thikana* is in the south of Mewar consisting of 110 villages. It was held by the Sarangdevot sub-clan of the Sisodias. The heads were designated as Rawat. Kanor is situated about 60.8 km. east by south-east of Udaipur city. In addition to the wells attached to the *Kamalwala Talab*, a large number of wells were constructed which were spread over their ancestral territory. Some of the evidence on it is compiled

by J.K. Ojha in *Kanor Thikaner ke Patte Parwane*, Udaipur: Pratap Shodh Sansthan, 1996, pp. 1–38.

25. Sirvis were prominent peasants in Marwar. They were considered the best cultivators. It is said that wherever they settled they transformed ordinary land into a fertile tract. In Marwar they were settled in the fertile area of the state, i.e. Sajat, Jaitraran and Jodhpur. The name *Sirvi* is derived from the word *sir*, meaning cultivation. When the Sirvis' fame reached the ears of the rulers of Mewar they invited them to settle in their territory. The Sirvis were granted wells and lands which show the interest of the Maharanas in irrigation and agriculture.
26. The local informants were Shri Manohar Singh, Dhan Singh and Abdul Ghafar, son of Sakur. They reported that both the step well as wells as ordering well are used to water surrounding fields. *Rahat* is still being employed to lift water from them. Both the harvests are grown. Wheat, maize and sugar cane are the main produce here. The water-table is quite near the surface because of the Jana Sagar.
27. Sita Ram Lalas, *Rajasthani Sabad Kos*, I, Jodhpur: Rajasthani Shodh Sansthan, p. 506.
28. 'Mewar Rajya men Mandalgarh Pargana-ke-Gaon-ki Vigat', op. cit., p. 22.
29. *Parwana* of Maharana Sangram Singh of VS 1768, Bilara collection, N.S.S. Sitamau.
30. Erskine, op. cit., p. 47.
31. James Tod, op. cit., p. 126.
32. Shyamal Das writes that in real terms rivers irrigate a major part of Mewar because water percolates from these rivers and remains near the surface which is called *seja* in local parlance. Therefore wells were excavated on both the sides. Obviously water-lifting device employed in *sejas* was *dhenkli*, cf., *Vir Vinod*, I, p. 118.
33. Qaisar classifies water-lifting devices into two categories: (a) intermittent or discontinuous water supply devices; and (b) a continuous supply system. He puts rope-bucket technique in Type A which was not used by any mechanical aid. See 'Agricultural Technology Depicted in Mughal Paintings', in *ITINERARIO*, vol. XVI, no. 2, Leiden: Grafaria, 1992, p. 69.
34. *Vir Vinod*, I, p. 118.

35. For details see, Qaisar, 'Agricultural Technology', op. cit., pp. 69–73.
36. James Forbes, *Oriental Memoirs: A Narrative of Seventeen Years Residence in India*, I, Delhi: Gyan Publishing House, rpt. 1988, p. 252 as quoted by A.J. Qaisar in his paper 'Agricultural Technology', op. cit.
37. *Vir Vinod,* p. 118.
38. Qaisar, op. cit., p. 73.
39. Ibid., p. 74.

CHAPTER 8

Step Wells

Water has always been an essential for human beings inspiring them to obtain it by the hardest of efforts. They developed all kinds of well-knit methods to harvest and conserve different types of water, namely underground, surface and rain water. They developed methods best suited to the kind of water to be utilized and the geographical and topographical conditions. The civil engineers appear to have thought of developing a kind of structure which could utilize both underground as well as rain water. The step well seems to have been the fruit of this thought. The motive of recreation appears to have been a later addition. Subsequently these structures were decorated with various architectural features.

In medieval and pre-medieval times innumerable step wells were constructed near the inhabited areas, both in the mountains and plains alike. The erection of a large number of water structures in Mewar is an interesting example of the response of the people to the need to exploit favourable climatic and geographical conditions. The hydraulic engineers designed these structures in such a way that they could protect stored water from all kinds of impurities. Therefore they made narrow stairs going deep down in order to expose as little surface-area as possible to the sun and to provide large areas of shade in the various storeys and galleries of the step wells. The dense forests and numerous trees also provided shade and protection to these structures. Forest vegetation is found everywhere except in a few cases. Thus the climatic conditions were favourable for storing water for daily use and irrigation.

As we know groundwater hydrology is the science of occurrence, distribution and movement of water below the surface of earth. The largest available source of fresh water lies underground; the total groundwater potential is estimated to be one-third of capacity of oceans. The main source of groundwater is precipitation. A portion of rain falling on the earth's surface infiltrates the ground, travels down and when checked by an impervious layer to travel further down results in reservoirs of groundwater. The groundwater reservoir consists of water held in voids within a geological stratum.

Countless step wells are found all over Mewar. The *Vir Vinod* provides information about the major structures and their inscriptions. During our survey in the major part of the state, we found that nearly every village and town has at least one step well or *kund* or well or pond. Some of them are ordinary while many of them have interesting architectural features.

A large number of such structures are found in the vicinity of Udaipur and its surrounding localities. Prominent among these are: *Ghosundi Baori* near Bassi, Gogunda; *Takshak Baori* near Kailashpuri; *Rana Baori* near village Chirwa; *Vyas Baori* between Eklingji and village Delwara; *Raniji ki Baori* near Shahpura; *Chamna Baori* in Pur, *Sand Baori* and a number of other *baoris* in the village of Kankroli; *Asind Baori* and other *baoris* in Bajolia; *Panthi Baori* and *Trimukhi Baori.*[1] The city of Udaipur itself contains several step wells to cater to the needs of the urban people and for watering the gardens. All these structures always remained full of water because of the high water-table caused by the existence of large lakes and tanks in the vicinity such as Rang Sagar, Pichhola and Fateh Sagar. In the vicinity, of the city palace are three large gardens containing more than ten step wells, including *Shyamji ki Baori, Neelkanth ki Baori, Navlakha Baori* and *Sagas ki Baori.* They were used for watering these beautifully designed gardens. Besides these, every locality of the city had a step well mostly used for daily needs.[2]

Of the various terms for step wells, *vav* and *baori* are in common use in Rajasthan. They were used to differentiate these wells from an ordinary well (*kuwa* or *kund*). It appears that the term *vav* is derived from the Sanskrit term *vapi* or *vapika*.[3] The term *baori* or *bawari* has two components: *ba*, meaning water and *wadi* or *ori*, means enclosed corridor or steps. Basically it is a combination of two characteristics of two water structures: one well and two, a chamber with descending steps reaching up to the mouth of well. The former was for tapping underground water while the latter was for collecting rain water, so a well with an overhead structure attached with steps was styled as a step well. It becomes clear from this that the *baori* was a peculiar type of well-monument which was used for religious rituals, drinking and irrigation. However, the term *baoli* does not appear in the Rajasthani documents.

A step well consists of four major components: one, a vertical well; second, an enclosed chamber, sometimes multi-storeyed; third, a flight of steps from the top pavilion down to the water level; and fourth, intermediate platforms down to the water level.

We come across various types of step wells which were located at different places. Broadly, these can be classified into five categories according to their location: those connected to a temple or shrine; those located in a portion of a garden or orchard complex; those, within or at the edge of a village; those on highways and trade routes; and last, those in the middle of an agricultural field or on the fringe of an inhabited area or field.

The location of a step well is extremely important since it decides its function. The main purpose of a well monument is obviously to supply water. In Mewar, water is available in plenty during the rainy season. The south and south-west part of the region contain enormous water sources[4] which not only raise the water level of the surrounding areas but fill up all the over-ground waterbodies. But these dry up within a few months. So the step wells became all the more important because they were sunk deep into the earth to tap underground springs.

The position of the step well itself determined the purpose of its construction. They not only supplied water for domestic needs but also for watering gardens, orchards and agricultural fields. We are mainly concerned with two kinds of well-monuments: first, monuments used exclusively for irrigation which we may call irrigational structures and second, double-pronged utilitarian bodies, i.e. those which supplied water for irrigation and drinking and for personal needs.

The extant structures of step wells attest to their widespread use for irrigation in Mewar. In the desert state of Marwar, too, these structures were employed for watering fields. Nainsi records that such step wells were used for this purpose.[5] They were extensively used for irrigating the fields of *pargana* Pokaran, located in the middle of the rocky desert.[6] All the *baoris* must have been brick-lined, else they could not have lasted long, or supported the steps and the various floors.[7] In Mewar too, we come across numerous structures which were utilized for this purpose. The well-monument used for this purpose had a sluice at its rim to receive the drawn-up water which was then lead into a trough or big pond (*haud*). From there it was taken through channels (*nali*) running into different directions inside the fields or orchards.

Step wells for irrigation and drinking were also found in appreciable numbers. The monuments used for irrigation were generally constructed adjacent to villages surrounded by agricultural fields. There were step wells located outside settlements too, on trade routes, highways and route-junctions which were obviously meant for travellers and caravans. We have come across during our fieldwork many water-monuments located on the main or subordinate routes. The *Panthi Baori*, *Trimukhi Baori* and *Debari ki Baori* are on the Udaipur–Debari route which is connected with Chittor. Similarly, on the Udaipur–Nathdwara highway numerous well-monuments are located at Bhuwana, Amberi and Bedla. Three step wells are found on the Udaipur–Dungarpur route namely, *Govardhan Vilas ki Baori* and *Titardi ki Baori*.[8] These monuments catered to the needs of travellers and caravans besides that of the local inhabitants.

The purpose of the construction of step wells was ordinarily twofold: fulfilling utilitarian and social needs, including the spiritual needs of devotees. Many step wells were erected as a part of temple establishment. They were attached to a temple and used exclusively for religious purposes. This is confirmed by the inscriptions on the well-monuments. This also gets corroboration from the actual extant structures, such as *Sunder Vinayakji Baori*, *Mahakal ki Baori*, *Shyamji ki Baori* and *Satyanarayanji ki Baori*, which belong to this category.

Religious considerations and beliefs appear to have encouraged members of the royal family and wealthy people to erect such structures. This gets corroboration from innumerable inscriptions on the step wells and from literary texts. Ramrasdi, the wife of Maharana Raj Singh, built a step well in the valley of Debari which was a part of the temple complex. Its cost of construction came to Rs. 24,000.[9] Similarly, another wife of the Sisodia ruler, built a similar monument in Raj Nagar adjacent to a temple which cost her Rs. 30,000.[10] Besides these, many well-monuments were constructed by the members of elite families, such the *Dhaji ki Baori* and *Mehtaji ki Baori*. The use of the water of these wells for performing rituals in the temples appears to have been under the control of the owners. In addition to it, sometimes the water of these well-monuments could be utilized to irrigate the gardens attached to the shrine. Generally, we find gardens being maintained by the *purohits* of the temples in which flowers and fruits were grown for offering to the gods and goddesses.

The *jagirdar* of Badhnor, Dungarsi, took keen interest in water harvesting in his *jagir*. He had built a number of step wells for the benefit of his people.[11] A man named Tarachand, probably a businessman, erected a step well in the town of Sadri which was named after him as *Tarachand ki Vapi*.[12] Cenotaphs (*chhatris*) were built over this step well and it was probably adjacent to a temple known as Ranpur. It appears to have been used by washermen of the locality for washing clothes.[13] A large number of references attest to the construction of step wells by the members of the royal family, nobles, wealthy merchants and sometimes even by ordinary people. Wells dug out by ordinary men were generally meant for utilitarian purposes like irrigation or drinking. Interestingly, we have also come across some step wells which were built by different caste panchayats as is evident from their nomenclature. These are *Nai ki Baori* (barber's well) in Udaipur, *Kunjaro ki Baori* (vegetable growers' well) in village Bari located behind *Bari ka Talab* or Janade Dam and *Mochiyon ki Baori* (leather workers' step well) in Shahpura.

The caste-designated structures indicate the involvement of caste panchayats in their construction. Though the absence of any written evidence forbids us to say more about the involvement of the panchayats in the construction of these structures, it would have been interesting to know how the finances for the construction were organized, how the caste-members were allowed to use its water and what they had to pay towards its construction. These questions could be answered only after getting more evidence.

II

The multitude of step wells form an important part of the building activities of Mewar in the pre-medieval and medieval times. The varieties of well-monuments clearly reflect the development of architectural and sculptural styles. The earliest step wells built of huge stone-blocks belong to the seventh century. The styles of the structures appears to change gradually.

Broadly three kinds of well-monuments are found in Mewar: one, single-sided steps with a *mandap* (platform) in the front or middle; second, staircases on two sides with the *mandap* in the middle; and third, steps in three directions with a central *mandap*. The *mandaps* of all wells were raised on pillars. We have also come across structures, designated as *kund* not *baori* which contain staircase in all the four directions. The historians of art and architecture express surprise over the majestic styles of these monuments and equate them with temples and assign them the status of pilgrim centres.[14] Raj Shekhar Vyas writes that temples were assumed to be the residence of gods and therefore their architecture developed accordingly in India. Similar status was assigned to step wells and *kunds* and they developed on the lines of temple architecture. Vyas is of the opinion that the well-monuments of

Mewar were strictly within the parameters prescribed by the Vastu Shastras (architectural treatises).[15] It is true that we find magnificent and highly stylized *baoris* in different parts of the state but they differ in styles. Their stylistic idiom appears to have changed with the passage of time, as in the case of Gujarat.[16] The historian Dhaky has classified the western Indian style into four groupings and coined new terms for them: Saurastra style, Maha-Maru style, Maha Gurjara style and Maru-Gurjara style.[17] We will not make any attempt here to classify the step wells of various styles found in different areas of Mewar. Our prime concern is to initiate a study of the structures for further in-depth research on the subject. This would certainly bring interesting results and throw light on the stylistic idioms of the step wells of the Mewar region.

Trimukhi Baori

The real significance of a step well lies in the functions it performed. Almost all these structures were built for a specific purpose such as watering agricultural fields or pleasure gardens or orchards, for providing water for drinking and for serving religious monuments. Generally, well-monuments constructed in the middle of agricultural fields had simple architecture. They were used for watering fields, for drinking purposes and for meeting other social needs. Wells erected by members of the royal family exhibited architecture according to the status of the builder. The *Trimukhi Baori* is one such monument which was raised by the wife of Maharana Raj Singh at a cost of Rs. 24,000.[18]

This *baori* is located in the heart of the village Debari and is surrounded by agricultural fields. The name of the structure, Trimukhi is self-explanatory—containing three entrances. Every passage leads to the mouth of the well through a flight of steps. Each side is divided into three parts and one section called *deri* or *kut* contains nine stairs which totals up to 30 (27 stairs + 3 *deri*) steps in one direction. The total of stairs and *deris* on all three sides comes to ninety. All the stairs are of equal measurement (.30 cm). The entrances are located in the south, the east and the west. The well is on the northern side where a platform has been raised for installing the water-lifting device.

The length of the waterbody from east to west comes to 60 m. while the breadth is 39 m. The length is higher than the breadth because of the absence of an entrance on the northern side. The vertical well is square in size measuring 6.45 × 6.45 m. On the northern wall a platform was erected to lift water from the well for irrigating fields located in the same direction. Apparently the *charas* seems to have been operated on it.[19] The ground plan of the structure is prepared based on survey (Figure 8.1).

Figure 8.2 gives a general interior view of the *baori* which seems to be a very large complex comprising three sections of the stepped water tank—two rectangular ones flanking the central one which is square in plan. The three sections are separated from each other by means of wide bridges, the lower portions of which contain trabeated large openings through which the three water chambers are interconnected with one another. The entrances are supported by multifaceted pilasters at the corners and *chhajjas* above. The *baori* is a double-storeyed structure. The upper portion above the bridges is in the form of pillared galleries with flat roofs and *chhajjas* supported by brackets. The pillars are square at the base and on the top while their shafts multi-faceted in the middle. The two sides of the upper storey have rooms with intervening pillared verandahs. The rooms overhanging the water tanks below are provided with beautiful windows for ventilation. The water tanks at the two entrance ends have plain stone steps descending to the water level while the middle tank has a beautiful arrangement of double steps on all the four sides of the square well see Figure 8.3.

Figure 8.3 shows steps descending from the ground level to the well through a bridged gateway. The gateway has *chhajjas* on both its sides which are supported by beautifully-carved bracket capitals resting on multi-faceted pilasters attached to the corners of doorways. On either side of the gateway are niches. Above the bridges are pillared galleries which connect the chambers built along the two sides of the well. The well, which is square in plan, has two stages of descending plinth with angular formation at

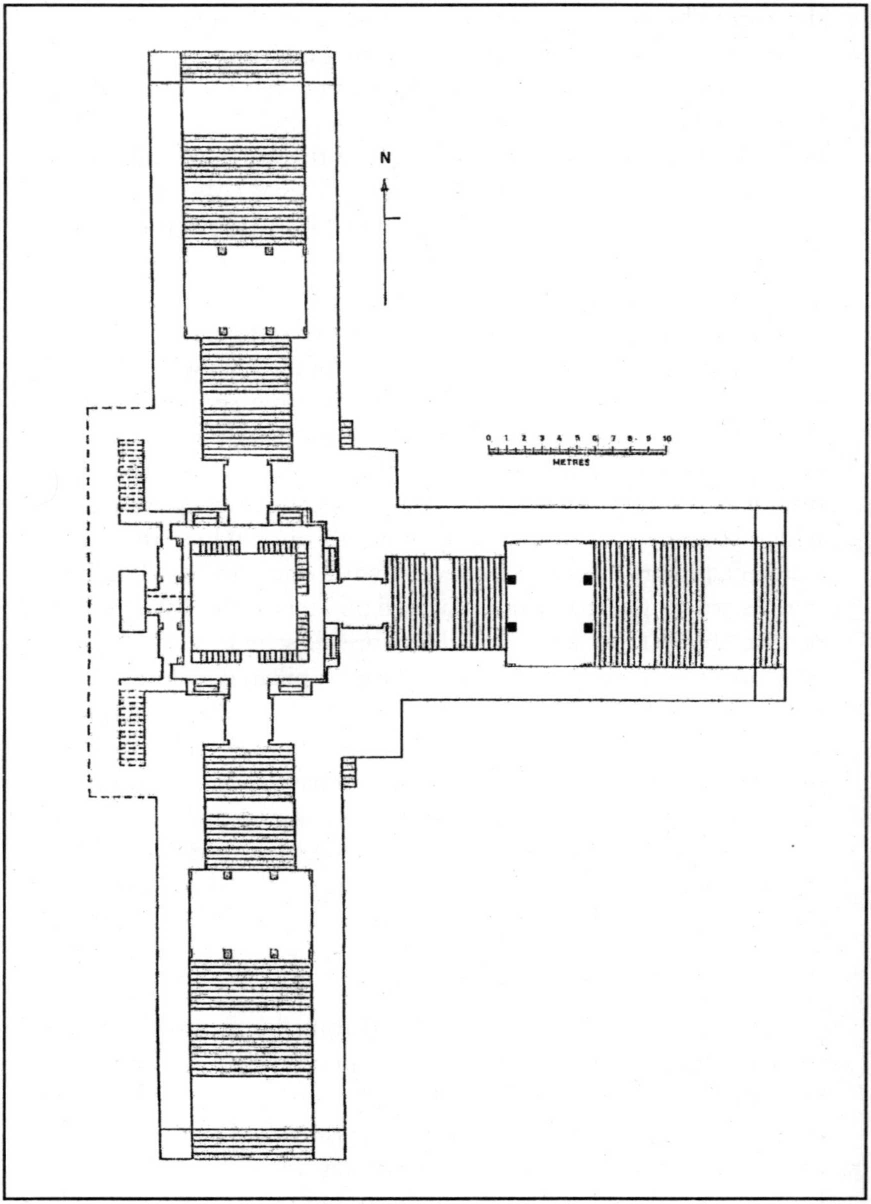

Figure 8.1: Ground Plan of *Trimukhi Baori*

Figure 8.2: Interior View of the *Trimukhi Baori*

Figure 8.3: Steps Leading to the Well

corners. A *chhajja* resting on beautifully-carved brackets runs above with similar angular formation corresponding to the angular plinth below.

The structure of the *baori* is magnificent. Figure 8.4 shows the details of the pillars in the upper storey which at the top terminated in four-sided bracket supports. At each ascending stage the shafts have beautifully carved patterns of different varieties. There is a pillared balcony overhanging central water tank. The lower portion of the balcony is a massive wall rising above the water tank. It has three projecting pillared windows with small *chhajjas* each crowned by a triangular floral pattern. The central one is slightly wider than the side ones. Above the central window are three lotus medallions. The upper portion of the wall is a three-tier formation of a beautifully carved parapet and mouldings. The parapet is decorated with a series of carved lotus medallions. The upper storey has a chamber with an entrance in the middle and two flanking ornamented niches on the side walls. The chamber is fronted by a pillared balcony supported by bracket capitals. The parapet of the balcony has a small projection in the middle which is supported by four beautifully carved brackets. The trabeated entrance of the chamber has beautiful bracket supports resting on nook-shafts.

Figure 8.5 shows the lateral stairs in four stages on two sides to reach the water level. These were specifically constructed to fetch sub-soil water of the well when its water level went down exceptionally. The entire structure of the step well remains full of water not because of sub-soil water but due to the water accumulated in the rainy season. The rain water comes into the well through showers over the structure. This raises the level of the water in the well-monument.

Panthi Baori

The step well located in village Bherawas or Berwas is known as *Panthi Baori*. The village is on the Udaipur–Chittor highway and falls in the municipal area of Udaipur.[20] This well-monument was

Figure 8.4: View of the Upper Storey of the *Trimukhi Baori*

Figure 8.5: View of the Lateral Stairs, Seen Through the Entrance

built by Pancholi Fateh Chand, the *pradhan* or premier of Maharana Raj Singh.[21] It was on the road known as *Nanda Panth*. Therefore, the name is probably derived from the word *panthi*, meaning traveller. A graffiti of VS 1743/AD 1686 tells us that the monument was raised in VS 1725/AD 1668. This writing was engraved by a stone-cutter (*silawat*) named Mansi on the corner of the niche in which an official inscription is installed. It appears that eighteen years after the construction of the *baori* a stone-cutter named Mansi might have passed through this way who recorded the year of its erection and the name of the engineer (*gazdhar*). The planner and designer of this structure was Kamal Hasan, son of Firoz. The engraver had recorded its size as 153.7 m.[22]

The inscription on the monument states that its builder's wish to earn name and fame led him to raise this step well. It was excavated in VS 1725/AD 1668 and a *sarai* was built and a garden laid.[23] It seems to have been a compact complex built on a trade route and was meant for the convenience of the travellers. Besides catering to the daily needs of the residents of the *sarai* as well as the villagers, the step well could irrigate a nearby garden spread over 13 *bighas*.[24] Once Maharana Raj Singh visited this site en route to the Raj Samand dyke in VS 1730/AD 1673.[25]

The structure, which appears to have been one-storey apartment, is in a dilapidated condition. We could not find any remains of the *sarai* and the garden seems to have disappeared due to neglect. All we could trace was a well located towards the south and an entrance on the east. It is a rectangular monument and the steps in the descending order start at the level of the surface. There are nine stairs in a row and then a broad platform (*deri* or *kut*) followed by another set of nine stairs of equal size. With the help of 18 stairs plus one platform one can reach an arched entrance measuring 3.00 m. Then there is a level platform up to the corner of the well.

The well of the *baori* is located in the southern direction. It has a diameter of 6.50 m. The well is deep and measures 18.3 m. from the top edge of its base. Steps are provided to reach into the well, probably to clean the silt from the bottom. At the back of the structure is a parapet which was used to lift water from the well. The *rahat* was probably employed for this purpose for watering the garden. The length of the structure is 35 m. and the breadth comes to 11 m. For ground plan of the *baori*, see Figure 8.6.

In the middle of the *baori* is an arched doorway which is surmounted by a pillared *chhatri*. The entire structure is built of red sandstone. The front view of the doorway (Figure 8.7) provides some interesting architectural information. It has a pointed semicircular recessed arched entrance which rests on side door jambs. The recessed arch is enclosed within a rectangular frame. The spandrels contain beautifully carved lotus medallions—one on either side—while the rest of the façade is plain. The entrance is flanked by two small stone seats (*asanas*) in the tradition of temple architecture, the top of which is decorated with a series of triangular *kanguras* in the form of a fringe of lotus buds. The same pattern is repeated on the projecting upper platform above the four walls of the *baori*.

As pointed out earlier, the archway is superimposed by a *chhatri*, again built of sandstone. The *chhatri* is built above a low platform, the four sides of which are slightly projected and decorated with the same type of *kanguras* as found below on the projecting platforms of the walled enclosure as well as the side *asanas*.

The *chhatri*, as usual, is an open-pillared pavilion composed of four-arched openings. The arches are multi-foliated comprising nine cusps which remind of the cusped arches of Shah Jahan's period. The shafts of the pillars are square up to the middle and then turn into an octagonal topped by simply carved cushioned capitals. The lower-middle square part of the shaft is interestingly divided into three diminishing stages. The *chhatri* is protected by curved projecting eaves (*chhajjas*) above the arched openings. Those curved eaves again resemble to the canopy of the emperor's throne in the *Diwan-i Am* of the Red Fort at Delhi. Above the *chhajja*, the pavilion is roofed by a vault after the Bengali fashion (or Bengali roof). The four sides of the vault are decorated with a panel of merlons supported by a moulding. The vaulted roof is pilastered and is crowned by *padmakosha* (lotus petals) and a pair of *amalakas* (melon like element) above (Figure 8.8).

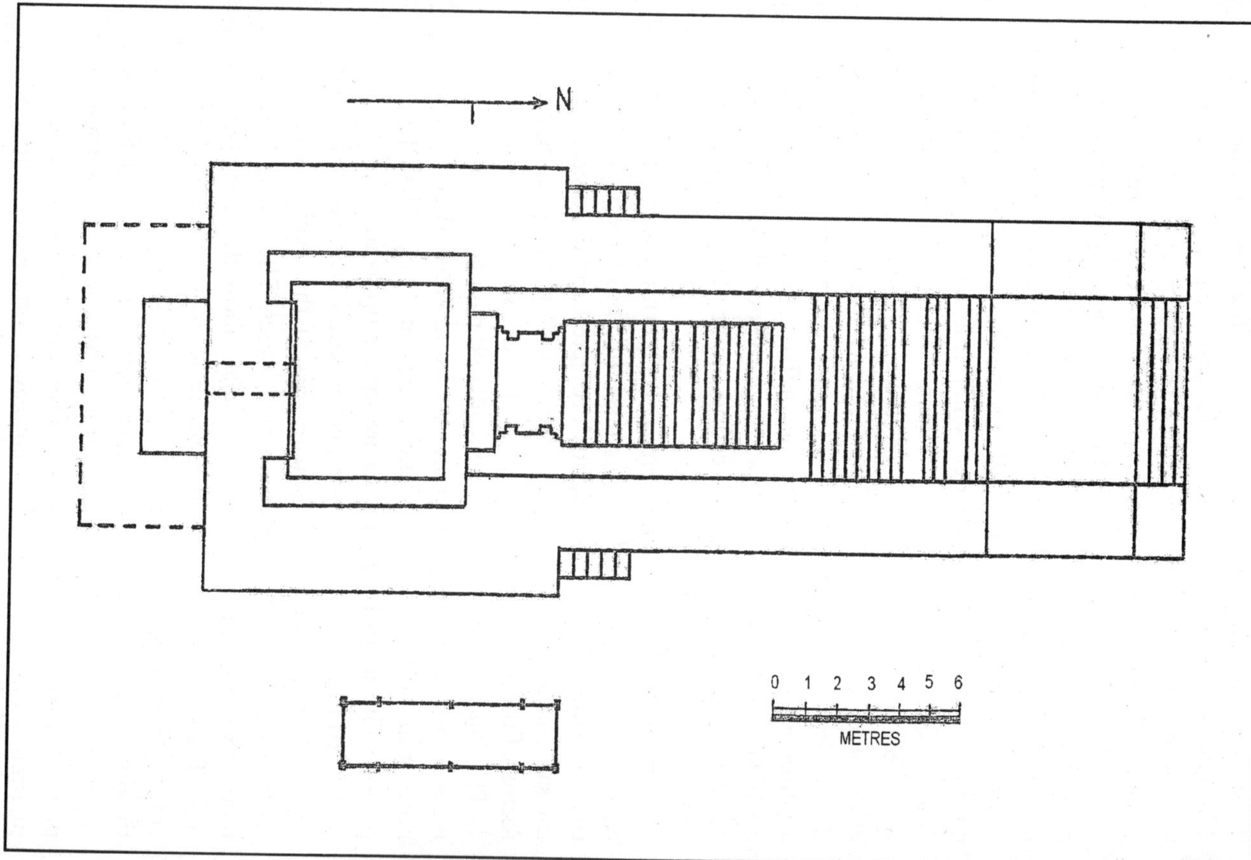

Figure 8.6: Ground Plan of the *Panthi Baori*

Figure 8.7: Front View of the *Panthi Baori*

Figure 8.8: View of Vaulted Roof and Lotus Petals

The rear view of the archway has a similar formation. A deep rectangular chamber is enclosed by huge walls on the north and south sides; on the east are steps leading down to the said archway on the west. The side walls along the steps are low walls in the form of two narrow platforms flanking the archway as they end with its quoins. The slightly projecting coping stone-slab of the two side platforms (low walls) is decorated with similar lotus buds (*kanguras*) as discussed above. The archway also has similar lotus medallions in the spandrels while the crown of the arch is decorated with a lotus flower.

Sundar Vav at Village Parda

A well-monument was built in the village Parda by a lady named Sunder who was a *patur* (sub-wife) of Maharana Raj Singh. She had the status of a *Khawasan*.[26] She undertook a plan to build a *baori*[27] near Udaipur. She selected the site at the village Parda located 4 km. to the east of the capital. The monument came up in VS 1717/AD 1660. On the occasion of the *pratishtha* ceremony, the Sisodia chief gave 75 *bighas* in charity in village Bhawana to Vyas Govind Ram and Balbhadra.[28]

The purpose behind the construction of this *vav* seems to have been twofold: one, incessant supply of drinking water for the residents of the locality and second, to irrigate the fields adjacent to it. This monument is still in good condition and provides water as in earlier times. It looks impressive from the architectural point of view too. The alignment of the structure runs from north to south. It is a single-storey compound with the well on the northern side and the entrance on the southern side. It has a magnificent arched gateway which is still intact and in good shape. A rectangular platform is provided which ends at the mouth of the well. The steps start from the entrance punctuated after seven stairs by a *kut* or broad step. Only 14 steps with two *kuts* lead to the well. On the three sides of the well there are three multi-foliated arches. In order to descend into the well staircases are provided on all the four sides. At the rear are two pillars above a ring to lift water, probably with the help of the *charas* for watering the nearby fields. The beauty of the structure lies in the dressed stones which are finely joined with lime-mortar.

At the centre of the entrance gate is a statue of Eklingji in a niche specifically made for it. Inside the structure on the first *kut* after the first set of seven stairs are two niches for placing the statues which seem to have been removed by vandals. In the front wall of the well is a multi-foliated statue still adorning a niche. This well is still in use for domestic purposes while its use for irrigating the fields has been abandoned.

Step Well at Bhawana

A step well exists in village Bhawana located 2 km. north of Udaipur.[29] This structure was erected by the mother of Vyas Govind Ram and Balbhadra on the land granted to them in charity by Maharana Raj Singh in 1660.[30] The village is near Eklingji and the well-monument is to the west of it. The *prashasti* on it says that the mother of the grantees had built a hydraulic structure along with a *sarai* which was known as *Lali ri Sarai* probably named after its builder. This latter structure was made during the time of Maharana Jagat Singh as evidenced by the inscription.[31] It means that a compact complex was erected which probably might have covered the entire 75 *bighas*. The intention of the lady behind the construction of the *baori* appears to have been philanthropic—she wanted to quench the thirst of travellers. By adding a *sarai* structure to it in her name its purpose was widened making utilitarian interest more prominent by catering to the needs of the caravans en route as well as ordinary travellers.

Step Well in the Garden at Udaipur and Rajnagar

Before ascending the throne, Maharana Raj Singh initiated the plan of a grand garden, *Sarvaritu Vilas* (pleasure garden for all seasons) in his princehood.[32] As a part of this project, a step well, *hauz* and a palace were erected.[33] The garden then constructed still exists but the palace has been converted into a hotel and the upper part of the step well is dismantled and the well is covered with stone slabs. But one can imagine from its remains the well-monument which once must have watered the pleasure garden of the royal family. It must have certainly contributed towards creating a healthy environment in the surrounding localities.

With the construction of the Raj Samand, the work of a step well was progressing simultaneously at the newly-founded city of Rajnagar. This hydraulic structure was initiated by the Rathor wife[34] of the Sisodia chief Raj Singh. Three days before the completion ceremony of the Raj Samand, the *pratishtha* of the *baori* was organized on *Magh Shukla* 7, VS 1732, i.e. on 23 January 1676.[35] The total cost of its construction came to Rs. 30,000.[36] The structure is intact but under the possession of a temple and remains locked, so it could not be surveyed. A glimpse of it showed that it is a quite a big waterbody made of white marble.

Rayno ki Baori

Bari is a small village in *tahsil* Girwa. It is 10 km north of Udaipur located in a hilly area.[37] This place is well identified with the *Bari ka Talab* or Jana Sagar Dam raised by Maharana Raj Singh in 1678.

The name *Rayno* is given to the step well. The attribution is probably derived from the caste designation of Bhil since a majority of the population consist of this tribe. It appears that before the construction of the Jana Sagar Dam this *baori* was excavated by the villagers or the leader of this tribe.

The well-monument is situated in the middle of this village and at the back of the dyke. It is an ordinary structure erected by villagers for watering a small patch of land. The step well is laid out in the north-south direction, with the well in the north and the entrance from the south. It has two pavilions, one runs parallel to the upper wall while the second is underground and directly attached to the well. The overall length from the beginning of the steps to the corner of the well comes to 19 m. The breadth on the ground level is 11 m. A set of six steps of 30 cm.

breadth and 2.40 m. length are made to descend from the ground to the first level. The lateral stairs of 30 cm. each on both sides are attached to reach a platform of 50 cm. breadth in size. Then another set of six stairs of 35 cm. descend to a platform of 70 cm. After that are three lateral steps in both directions to reach a broad platform of 70 cm. Again the sequence of five lateral stairs on both sides reaches down to the water level. This means it is a three-storeyed underground structure. The last pavilion is attached to the well. It is a simple structure made of undressed stones. There is no carving or decoration of any kind and is now in a dilapidated condition. For ground plan, see Figure 8.9.

The well had a diameter of 8 m. The inside wall is cut out in rock and above it some upper slabs with holes are projected towards the well to lift the water. This water body is located in the middle of a garden which clearly demonstrates its utilitarian purpose. It was mainly used to water the orchard. The same garden also has a well structure which was used for watering the plants. The existence of two water structures in a single orchard compound encourages us to conjecture about its ownership. It was a *khalisa* village and in all probability it might have belonged to a local *bhumia*, either from a Rajput clan or the Bhil community. The simplicity of the water-monument goes in favour of the latter category.

There are some stones on which two kinds of soldier-like figures are carved out. One set of soldiers is dressed in turbans with moustaches and holding swords in their hands and other have arrows on the bows. All these miniature statues are worshipped by the villagers, showing the connection of village with soldiery. There is a strong possibility that the Bhil *bhumia* had soldiers from his own race who had taken part in some battle.

Kunjaro ki Baori

Kunjaro ki Baori is located in the village Bari, on the fringe of an agricultural field. The Jana Sagar Dam, the *Rayno ki Baori* and two wells are all at a little distance away. The water level of the

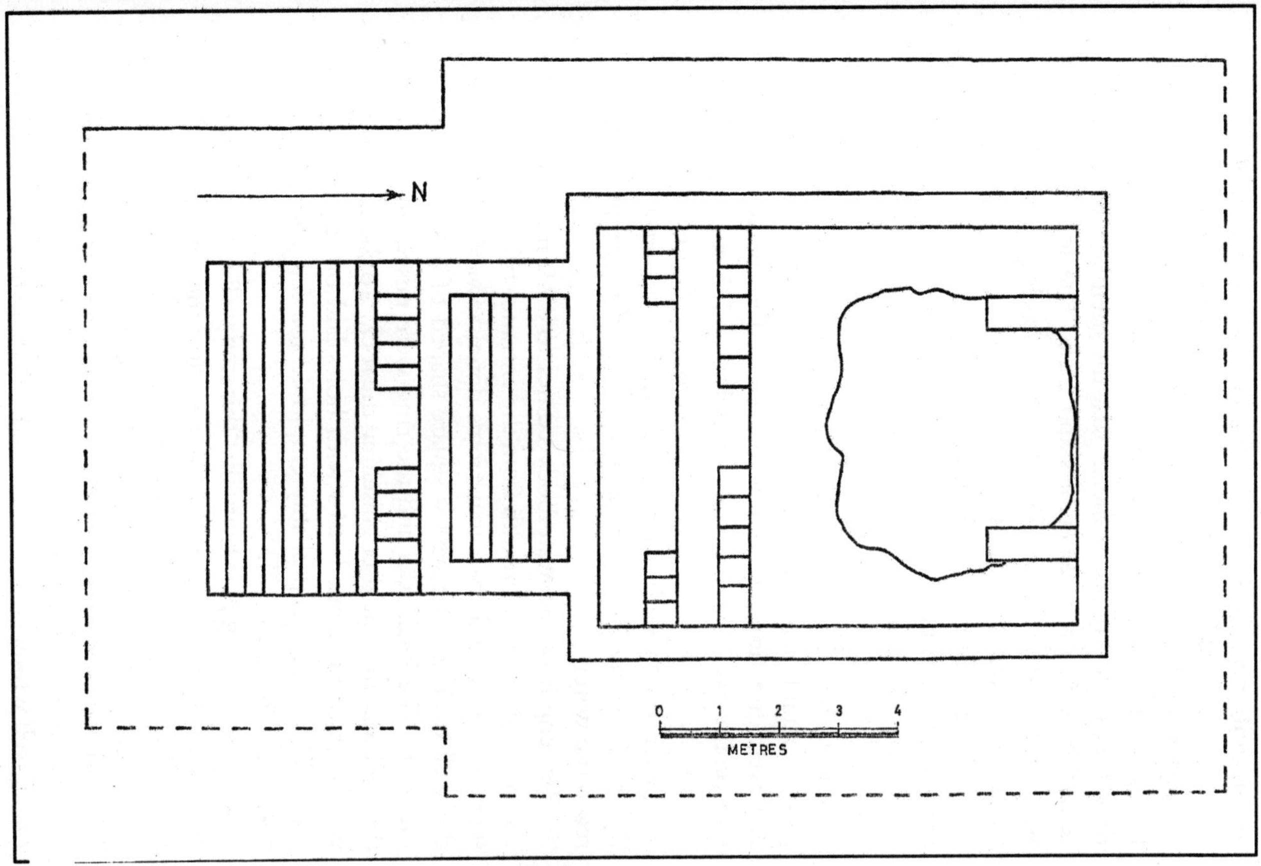

Figure 8.9: Ground Plan of *Rayno ki Baori*

area is quite near the surface because of the percolation of water of the canal from the dam which passes near the waterbody.

Kunjaro ki Baori derives its name from a Hindu caste known as Kunjara whose business is to sell fruits and vegetables. In fact, they are green grocers. It means that the members of this caste settled here to grow vegetables and fruits. The location of the waterbody on the fringe of a vast stretch of fields indicates that it was used for watering this tract.

This step well is a simple and unpretentious monument. It is made of stone-rubbles and minimum cementing material is used in binding the stones on three sides; the southern direction has a wall of stones.

In fact, the *baori* matches the social and economic status of the petty vegetable sellers. The alignment of the structure runs in the east-west direction. The entrance is located in the west and the well on the east. There are nine straight steps 35 cm. in size. The size of the platform at the entrance is 1 m followed by eight steps of equal size (80 cm) while the last stair narrows down by 20 cm. Then five lateral steps in one direction are provided to reach the water.

The length of the waterbody comes to 19.5 m. A platform encircles the well. Its breadth from the entrance till the end of the stairs is 8.5 m. and then it narrows down towards the back of the well to 5 m. The well is square shaped of 6 × 6 m. in dimension. The back wall and the ring were raised on rock and the well was dug out after chiselling the rock. To the south of the well are two horizontal slabs with holes laid on its ring to lift water. The *rahat* was probably employed for watering the adjacent fields (Figure 8.10).

There is a strong possibility that this step well is older than the *Bari ka Talab.* It was probably dug out after the founding of the city of Udaipur. As the urban demand for vegetables and fruits rose, then the members of this caste settled in this Bhil village. Thus this well-monument was in fact an ordinary irrigation structure watering a vast tract around it. The local people also call it a *rahat* well which clearly indicates the employment of this apparatus for drawing water.

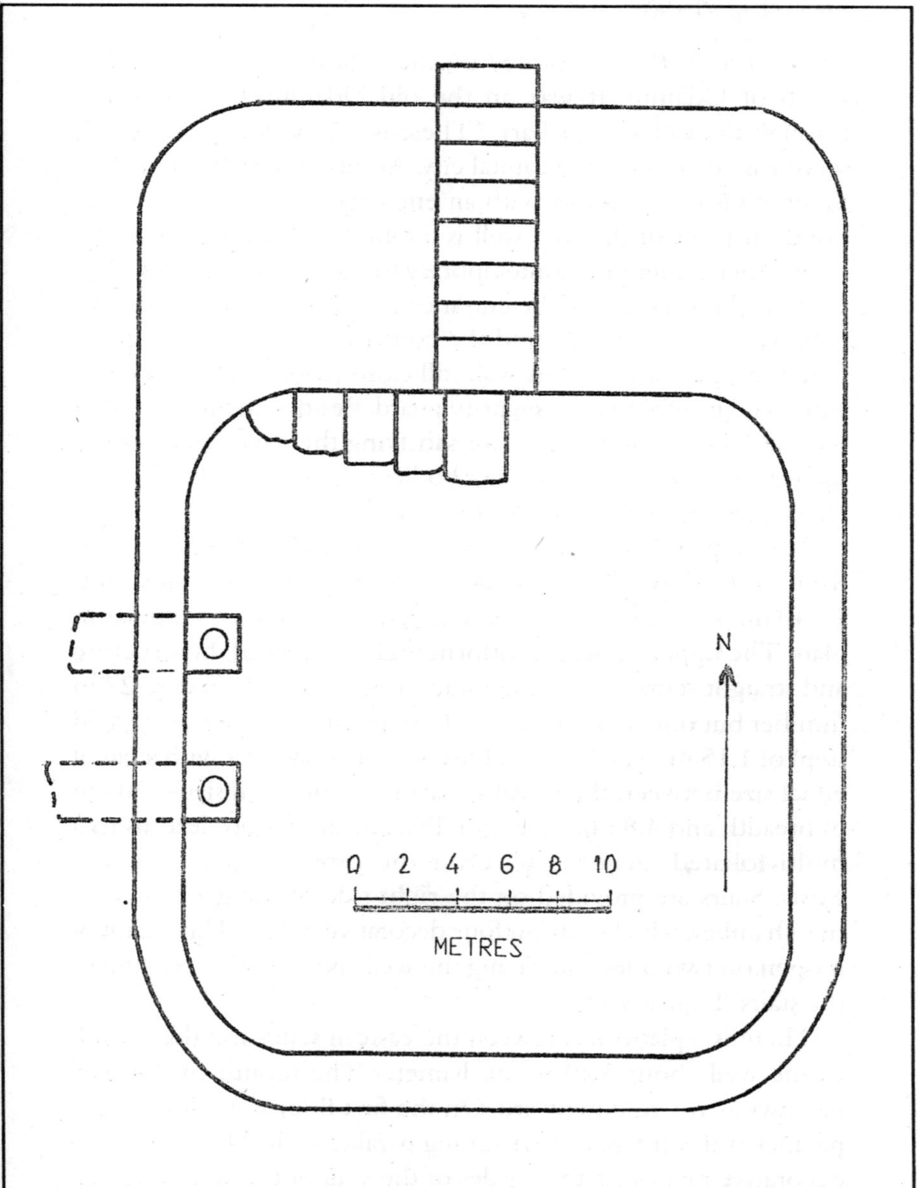

Figure 8.10: Ground Plan of *Kunjaro ki Baori*

Debari ki Baori

The *Debari ki Baori* is located on the national highway 10 km. north of Udaipur. It was on the old Udaipur–Chittor route through the valley of Debari.[38] There is a narrow passage which served as an entry to the capital city. At this point Maharana Raj Singh constructed a fort with an entry gate, a *baori* and a small pond. In front of the step well is a temple which was probably built either earlier or is contemporary to these constructions. The safety and protection of the entrance to Udaipur appears to have been the sole motive behind the construction of this complex. The front part and the fort wall still exists along with a big entry gate[39] and provision for a security guard. Besides chambers, a step well and a pond were made for satisfying the daily needs of the security personnel. These water bodies also served ordinary travellers and caravans passing by this route.

The step well is parallel to the fort wall. Its alignment runs from north to south. Its length comes to 37 m. while breadth is 15.40 m. The entire structure is underground and rectangular in plan. The upper floor is a platform on both sides of the structure and straight stairs lead to the underground floor. Stairs are 27 in number but one set of steps is of 11 stairs after a *kut* or *deri* (broad step) of 1.15 m. The baori had two sets of eight stairs with a *kut* of equal size between them. All the stairs are of equal size—30 cm in breadth and 4.85 m in length. Thus in all 29 steps lead us to a multi-foliated arched gate. Over the gate an open chamber exists. Stairs are provided on the right side of the gate to go to this chamber, which rests on four decorative pillars. The chamber is open on two sides, one facing the well and the other leading to the stairs (Figure 8.11).

There is a platform between the eastern stairs and the mouth of the well about 3.40 m. in diameter. The mouth of the well narrows as we go downward. On the first floor is a *chhajja* and a parapet wall on two sides running parallel to it. There are nine decorative niches on three sides of the wall of the well followed by one niche on each side below. A set of stairs in four stages is

Figure 8.11: View of the *Debari ki Baori*

provided on the right side of the wall to reach the water. Three platforms with steps on both the sides have been erected. In the last, steps on both the corners were joined to reach the maximum depth of the well to tap the sub-soil water (Figure 8.12).

The well on the south direction is octagonal in shape. Water was lifted from this side with the help of a *charas* and was ordinarily used for drinking purposes and for catering to the daily needs of the residents and travellers alike.

Masons' marks are found on the body of the structure. These probably represent either the marks of an individual artisan or a guild's marks.[40] It can be said that the *baori* is a structure with simple architecture having no decorative figures of any kind and erected for the use of security forces as well as for travellers.

Uncha Vav at Eklingji

Uncha Vav is located on the corner of the main road and in front of the Eklingji Temple. It probably derives its name from a small shrine standing over its wall. The inscription in the niche on the left side gives its year of construction as VS 1732/AD 1675.[41] Its designer was Sutradhar Kesavdas Teja.

This well-monument was part of a complex which contains a Shiva temple and a garden. The structure is laid in the east-west direction with the well in the west and the entrance on the east. The entrance starts from the ground floor and there are 14 stairs of 3 m leading to the well underneath. Below this is a broad step with three stairs on all the three sides followed by a broad platform in three directions attached with lateral stairs to reach the water level. In fact, the well is located on the western side and is covered by a stand alone rectangular compartment which has two openings in a square tank. The purpose behind the construction of this tank was religious. Lateral stairs are attached to the side of temple to take up water for rituals.

The purpose behind leaving two openings on the wall of the well towards the tank was to supply water at two levels. At the same time they also demonstrate the level of water in the well.

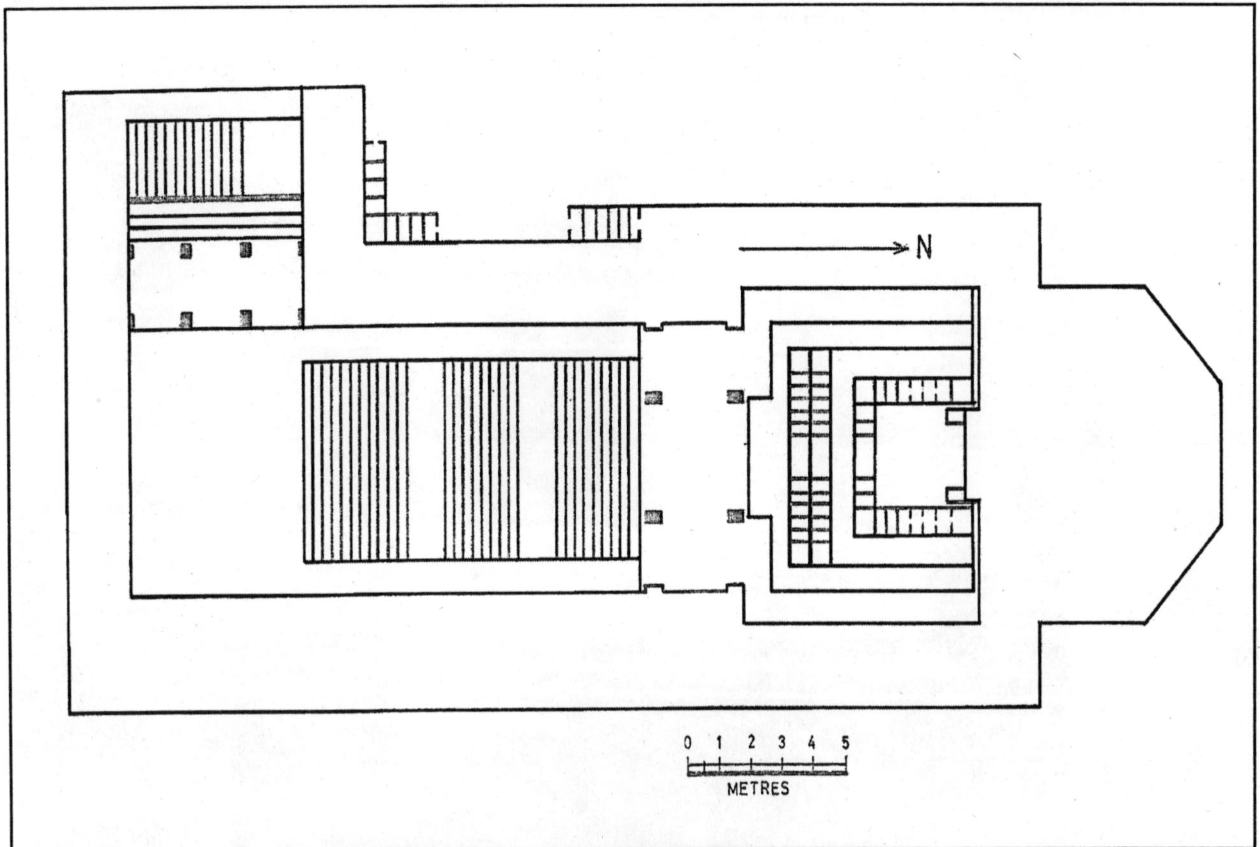

Figure 8.12: Ground Plan of the *Debari ki Baori*

Figure 8.13: View of the *Uncha Vav* at Eklingji

On the northern and the western direction are big gardens which were regularly watered by this well. A structure on the ring of the well was erected where a *charas* was operated to lift water with animal power.

Dressed stone slabs of variable sizes have been used here. Four niches at the entrance of the staircase are empty except one with a Sanskrit inscription. There are niches under the lateral stairs too. Separate stairs were constructed, attached to the temple (Figure 8.13).

The engineer of this structure was Sutradhar Kesavdas Teja who had also prepared its plan. Joshi Ram Kisan Mono and Piru were also associated with the construction of the waterbody. In fact, the characteristics of the step well and a *kund* have been incorporated in this *baori* and it is a perfect example of an utilitarian structure.

Thus step wells were waterbodies extracting subterranean water and harvesting rain water and served manifold purposes including irrigation. Numerous such structures were raised for watering the sprawling gardens and vast tracts of agricultural fields in addition to quenching the thirst of the people and providing water for rituals.

NOTES

1. A very large number of step wells are still extant in Mewar. It is difficult to catalogue them all but we have surveyed more then 40 *baoris* located in different parts of the state. With the help of the draftsman and photographer of our Centre we have documented them. Important among these are: (1) *Debari ki Baori*, (2) *Trimukhi Baori*, (3) *Rayno ki Baori*, (4) *Kunjaro ki Baori*, (5) *Adeswara Baori*, (6) *Jhanjhar Baori*, (7) *Sarvaritu Vilas*, (8) *Panthi Baori*, (9) *Mahakal ki Baori*, (10) *Mukarwa Baori*, (11) *Shyamji ki Baori*, (12) *Nai ki Baori*, (13) *Govardhan Vilas ki Baori*, (14) *Tatardi ki Baori* and (15) *Khanji Pir ki Baori*.

   Interestingly, the city of Udaipur alone has more than a hundred step wells. These played a positive role in 1987–8 when the city faced its worst water crisis owing to widespread drought. This shows that these medieval structures are still useful and can be activated in modern times.

2. The contemporary poets depict the beauty of the city. A poet named Maan writes that there are numerous gardens all around the city. Step wells, lakes and other reservoirs provide plenty of water. Cf. *Raj Vilas*, p. 32; for description of the *Sarvaritu Vilas Bagh* and *Vapika* (step well) built by Raj Singh see ibid., p. 57; Hem, a Jain poet, records the existence of a *vapi* (step well) and a *kup* (well) in the city, 'Medpat Varnan Prashasti', in *Maru Bharti*, no. 2, July 1954, p. 49.
3. The term *vapi* or *vapika* or *bapika* was also employed by the Rajasthani poets. See, *Raj Vilas*, p. 57; Agam Sagar, 'Sadri-ri-Gazal' in *Maru Bharti*, no. 4, January 1981, p. 55.
4. Nainsi records that the mountains located in the south-west contain enormous water resources, *Khyat*, pp. 40–5.
5. B.L. Bhadani, *Peasants, Artisans and Entrepreneurs*, pp. 52-3.
6. In the *qasba* of Pokran alone there were 25 step wells which were irrigating fields. All are brick-lined (*pakki*). It is interesting to note that the Persian-wheel was operated on these structures. On one structure two water lifting machines were employed for irrigating fields which shows the abundant availability of water. Wheat was grown in the locality. It appears from the evidence that these step wells were constructed by the rulers as well as by the ordinary people. The step well named *Mehavay* was built by a lower-caste person belonging to the *kunbhar* or potter's caste. The Rathor rulers also erected well-monuments and granted them in charity to temples and Brahmans. This demonstrates a joint venture between the ruler and the public (cf. Nainsi, *Vigat,* II, pp. 312–15).
7. Bhadani, op. cit., p. 53.
8. Raj Shekhar Vyas gives details of the *baoris* on various routes see, *Mewar ki Kala Aur Sthapatya*, p. 222.
9. Ramrasdi was the daughter of the Panwar Rajput who constructed this step well as a part of the temple. Secondly, the valley of Debari was on the Udaipur–Chittor route, therefore, it also served travellers (cf. *Rajprashasti,* p. 140).
10. The wife of Maharana Raj Singh, daughter of Rathor Rup Singh, known as Jodhpuri built this step well to serve the devotees of the temple at Raj Nagar (cf. ibid., pp. 146–7).

11. Dungarsi Chauhan was the *jagirdar* under Rana Sanga. The Sisodia chief assigned him Badnor in *jagir* in the beginning of the sixteenth century. Since *jagirs* in Mewar were assigned on an almost permanent basis, the assignees were interested in the development of their assigned territories. Chauhan Dungarsi constructed tanks, step wells and palaces in his *jagir* territory. Majority of the water-bodies might have been used for irrigating the fields (cf. Nainsi, *Khyat*, I, p. 119). It was close to the border of the British district of Merwara. The town is located about 153.6 km. north-east of Udaipur city. This *jagir* was assigned to the family members of the valiant Jaimal Mertiya who was killed in the siege of Chittor in 1567 (cf. Erskine, op. cit., pp. 90–1).
12. Sadri town is situated in 24° 23' N and 74° 43' E, about 105.6 km. east-south-east of Udaipur city. It was a flourishing town from the eighteenth century, as described by Agam Sagar in VS 1813/ AD 1756, but later on it declined. See Agam Sagar, 'Sadri-ri-Gazal' in *Maru Bharti,* year 28, pt. 4, January 1981, pp. 53–7.
13. Ibid., p. 55.
14. Rajshekhar Vyas, *Mewar ki Kala aur Sthapatya*, pp. 222–3; Jutta Jain Neubauer, *The Step Wells of Gujarat*, pp. xiii–xiv. Both these art historians emphasize this aspect of the step well. Jutta Jain writes about Gujarat and associates all these well-monuments with gods and mother-goddesses (cf. pp. 5–9).
15. Rajshekhar, op. cit., p. 223.
16. Gujarat is on the border of Mewar where the same kind of difference is noticed by Jutta Jain. She writes that:

    The Western Indian style does not display a congruous order in itself, it is convenient to divide it into various sub-groups. As appellations according to the royal dynasties were found insufficient for labeling the styles of temple architecture, it is suggested by some scholars that the naming of the various sub-groups according to their regional occurrence is more appropriate. The justification for such a classification is that certain stylistic idioms continued to survive even when dynasties changed. Moreover, the architectural styles often evolved independently of royal patronage. Many inscriptions and other literary evidences testify that ministers, merchants, members of religious communities, such as Jaina and Brahmana, guilds of craftsmen, and in many cases the wives of rich citizens sponsored the construction of religious and civic edifices. It is, therefore, thought more appropriate to use the regional term for denoting a certain idiom of style.Although regional terms are more appropriate, in certain cases it is indispensable to use dynastic names in order to classify a certain idiom of style more clearly, as its appearance can vary from period to period. These variations can be suitably determined according to the ruling dynasties in the various regions. For example, the Maru Gurjara style swept over an extensive area and prevailed for quite a long period of time, so that it is justified to use the chronology of the dynastic rulers for a closer determination of style-variations. Certain scholars, however, also use sub-regional names for these, like *Sapadalaksa* (for the Sambhar area) or *Medapata* (for the Mewar area). Op. cit., p. xiv.

17. M.A. Dhaky, 'The Genesis and Development of Maru-Gurjara Temple Architecture', in *Studies in Indian Temple Architecture*, ed. Pramod Chandra, Varanasi, 1975, pp. 114–65 as quoted by Jutta Neubauer, op. cit., p. xiv.
18. *Rajprashasti*, p. 140.
19. Two opinions about the water-lifting apparatus were expressed by the local people. One old man of 80 years said that the *rahat* was operated on it to draw water, which was carried out through the channel (or *nali*) which has disappeared now. The second recorded the use of the *charas*. But one fact which emerges from their evidences is that the step well was irrigating the surrounding fields.
20. The village Berwas, now called Bedwas, is located 8 km. north of the city of Udaipur. Formerly, it was a part of *pargana* Girwa whose estimated income was Rs. 400 (cf. *Maharana Raj Singh Pargana Bahi*, op. cit., p. 3).
21. *Prashasti* of Berwas *baori* in *Vir Vinod*, II, pt. 1, Appendix 2, pp. 381–3.
22. The graffiti left by the artisan Mansi goes like:
Sanwat 1725
Lakhat Gajdhar
Kamal Hese
Sut Firoj Do....
Gaj 252
Sudi 13 Gure sambat 1743

Silawat Mansi
Roysa Mitu
This date matches with the official inscription.

23. The work on the *baori* was completed on Monday, *Baisakh Shukla* 10, VS 1725.
24. Along with the well-monument, a *sarai* and a garden were constructed (cf. *Vir Vinod*, II, pt. 1, pp. 381–2).
25. Maharana Raj Singh was on his way to the Raj Samand Lake when he visited the step well and tested its water which was found to be good in taste (cf. ibid.).
26. The word *khawasan* appears to have its root in Persian. It is probably derived from the word *khas* meaning intimate. The feminine *khas* came to be known as *khawasan* and masculine as *khawas* in Rajasthan.
27. Two other wives of the same Maharana erected *baoris* at Raj Nagar and Debari. She was the third lady who took up this plan. This shows that a spirit of competition appears to have been working among the royal women.
28. *Vir Vinod*, II, pt. 1, p. 436.
29. Bhawana was in *pargana* Girwa (Udaipur) and still in the *tehsil* of the same name. Its estimated income was Rs. 1501 in the 1660s. Cf., *Maharana Raj Singh Pargana Bahi*, op. cit., p. 2. Its realization appears to have reached up to Rs. 300.
30. Maharana Raj Singh gave 75 *bighas* of land in charity to two brothers in 1660 on performing the *pratishtha* ceremony of a *baori*. See *Vir Vinod*, II, pt. 1, p. 436.
31. The inscription on the well-monument clearly mentions that the *baori* was constructed by the mother of the revenue grantee 'but the *sarai* was raised during the time of Jagat Singh, son of Raj Singh' (cf. Inscription text in ibid., p. 578). The translation of the text is as follows:

    The Maharana Raj Singh constructed a *Sunder Baori* at village Parda. On that occasion he gave 75 *bighas* of land in religious grant in VS 1717 to the Nagar Brahmans of Bisalnagar, Vyas Govindram and Vyas Balbhadra, son of Gopal. Their mother had built a *baori* and the *sarai* was constructed by the successors during the time of Jagat Singh, son of Raj Singh.

32. *Rajprashasti*, pp. 58–9.

33. Maan gives a description of the *Sarvritu Vilas*, a garden which was planned in a systematic way by Raj Singh. It was a large garden with a large step well for watering plants. Every kind of fruit plant was grown in it. Cf. *Raj Vilas*, p. 57; *Vir Vinod*, II, pt. 1, p. 443.
34. The name of the wife is not known but she was the daughter of Rathor Roop Singh of Krishangarh. She was better known as the Jodhpuri Rani (cf. *Rajprashasti*, pp. 146–7; *Vir Vinod*, II, pt. 1, p. 449).
35. *Rajprashasti*, pp. 146–7.
36. Ibid.; *Vir Vinod*, II, pt. 1, p. 449.
37. The village Bari Lohari was a part of *pargana* Girwa during the time of Maharana Raj Singh. Its estimated income was Rs. 201 (cf. *Maharana Raj Singh Pargana Bahi*, op. cit., p. 2).
38. *Vir Vinod*, II, pt. 1, p. 476.
39. An inscription is engraved on the wall of the gate which gives the date and year of fixing the wooden gate. This is being given in Roman:

    Maharajadhiraj
    Maharana Raj Singhji Aadeshatu
    Sawan Sudi 5
    VS 1731 Kimar Chadhwaya
    Likhtam
    Pancholi Nathu

40. Masons' marks are arrow-like, *damru*, sickle and so on.
41. The inscription is in Sanskrit and was engraved by Sutradhar Kesav Das Teja on *Magh Sudi* 13, VS 1732.

PART III

SURFACE WATER AND URBAN WATER MANAGEMENT

CHAPTER 9

Tank Irrigation

The importance of tanks for irrigation has been recognized since the earliest times.[1] The construction of the waterbody assumed religious as well as worldly significance for rulers, villagers and urban people alike. Archaeological survey reports and the gazetteers attest to the existence of at least one tank in most of the villages. It is believed that many of them were used for irrigation which is attested by Erskine's statement that of the 512.29 sq. km. of settlement, 102.4 sq. km. were irrigated by tanks and reservoirs.[2]

This confirms the significance of tanks among the many methods of artificial irrigation prevalent in medieval Mewar. This also gets supports from modern geographers.[3] The possibility of the construction of reservoirs, tanks and dams was more in the south than north Mewar. The former southern part of Mewar has a number of perennial streams while the northern region, specially the north-western part, possesses more porous soil with a high rate of evaporation which make surface reserve temporary. Therefore, this area is not well-suited for recharging ground water to improve reserve to be tapped by wells.[4] This explains the reason for the construction of major waterbodies in the south or in the south-west.

The Sisodia rulers who ruled over Mewar from *c*. AD 800 till 1947 developed an extensive system of reservoirs and tanks in their territory. This is evident from the available records as well as the extant structures. References of these waterbodies are replete in the early medieval inscriptions. For instance, inscriptions of AD 644 and 689 of the region mention two tanks constructed for watering the fields.[5]

In our survey we found sufficient number of these structures in Mewar whose contribution in water harvesting and conservation had been immense. They were also important for raising the water-table through percolation of water and also in raising water level of the surrounding wells. Therefore, efforts were made by all the concerned people to excavate *talabs* of varying sizes using the natural depression if one was there or to make one to tap rain water. The kind of waterbodies constructed depended on the resources of the persons involved in this task. The reservoirs dug out by the rulers or *jagirdars* or rich people are found to have stone work while the structures erected by the villagers were of mud. But the function of both the types of water monuments was the same.

Before proceeding further, it is necessary to make a distinction between a dam and a tank. In our sources two distinct terms are employed for dams and reservoirs. For the former body the terms *samand, sagar* or *sagararaj sarovar, pal*[6] and *setu*[7] are used while the latter are referred to as *talab, talav*[8] or *tadag.*[9] In this section, tanks or *talabs* have been described covering only those which were surveyed by the present author and which have been documented in the form of plans and photographs.

TANKS OF KANOR

Kanor is the *thikana* headquarters owned by the Sarangdevot, a sub-clan of the Sisodias.[10] This was assigned with a hereditary *patta* to Sarangdev in VS 1768/AD 1711. The town is situated in 24° 40' N and 74° E, about 65 km. to the south-east of Udaipur.

The town has five tanks, namely *Kamalwala Talab, Dhaneria Talab, Joshila Talab, Balera Talab* and *Kund Jharna* connected with each other. On the east and the south-east of the *Kamalwala Talab* are the *Dhaneria* and the *Balera Talab* respectively while the *Joshila Talab* and the *Kund Jharna* are in the south at a lower level. They are designed and planned in such a way that the overflow or surplus water of one tank fills the tank located below. Thus the hydraulic engineers created a web of tanks utilizing the geographical and topographical features of the area in an ingenious way. They continually erected embankment walls with mud and rubble or boulders to retain water coming from the hills located in the north-west. They constructed strong *bunds* of mud-rubble around the tanks with great care and skill. To provide additional strength, sometimes trees were planted by the side of *bunds*.

In the absence of any feeding canal connecting these water-bodies one may conjecture the overflow of water over the wall, though the danger of washing away of the embankment was always there.

The overflows of the *Kamalwala Talab* and the *Dhaneria Talab* go to the *Joshila Talab* located downwards while that of the *Balera Talab* fills the *Kund Jharna* and ultimately the combined flow of the five waterbodies falls in the Jai Samand in the south-west.

Kamalwala Talab

The local people named this waterbody as the *Kamalwala Talab* because of its lotus flowers but in our survey we also found many wells on the banks of the waterbody which give the impression of a lotus flower and that is why probably it was so designated in the early period (see Figure 9.1).

It is a beautiful tank, probably built by the *thikanedar* of Kanor with the help of the local people. Its length from the north to the south approximately comes to 250 m. while breadth from east to west measures about 200 m. On its bank wells were dug by peasants whose agricultural land was situated nearby, obviously for watering their fields. All wells are located in the north and east direction on the corner of the mud wall. The ownership of the well in the present times supports this assumption.[11]

Interestingly the wells located on the bank of the tank are called as *seja* wells which are generally dug on the banks of rivers. These wells receive water percolated from the tanks and thus they are fed from these tanks. All the wells located on the north and east had a small channel (*nali*) which carried water to the agricultural fields. A stone platform was erected to lift water from the well. A rope mark on the stone indicates the use of the *dhenkli*. The

Figure 9.1: Artist's view of the *Kamalwala Talab*, Kanor (Udaipur)

wooden frame attests this fact. The water thus lifted was diverted through the channels towards the fields (Figure 9.2 A & B).

The source of water is located in the north-west hills. The inlet is located in the north while the outlet is in the south. It seems that a man-made depression was made by digging at the site to retain the running water. The mud obtained by digging seems to have been utilized in raising the embankment wall. In addition to mud, rubble was used. Stone stairs were added at different intervals to reach the water level. To provide additional weight to the mud-rubble wall, *chhatris* (memorial cenotaphs) were erected on the northern side while a temple of Lord Shiva was raised on a stone platform over the wall (Figure 9.3).

It is difficult to determine the period of construction of this water monument in the absence of any written document but an inscription on the *chhatri*, though mutilated, gives us a clue. One of the stone writings mentions the year which is either VS 1709/AD 1652 or VS 1791/AD 1738. The latter date appears to be more correct because by that time the *pargana* was in the the possession of a *jagirdar* and the *chhatri* was probably built in memory of the first *thikanedar*. This suggests that the waterbody was built by the *thikana* of Kanor.

The *Dhaneria Talab* and the *Balera Talab* are located almost parallel to each other. The latter is to the south of the former. The former receives water from the northern direction and it discharges its surplus water from the outlet in the south. It appears that it feeds *Balera Talab* whose inlet is in the east. In addition, it also supplies water to the *Joshila Talab*, while that of the *Balera Talab* goes into the *Kund Jharna* through its outlet in the north. The latter waterbody appears to have been so designated because it receives many streams from all the tanks located upward (see Figure 9.4).

Mahurda (Mahuda) ka Talab

Mahurda or Mahuda is a village in *pargana*[12] Sadri located 14 km from Kanor. There is an enormous tank made of mud in the

A

B

Figure 9.2: Two Wells on the Corner of *Kamalwala Talab*

Figure 9.3: View of the Temple and Stairs

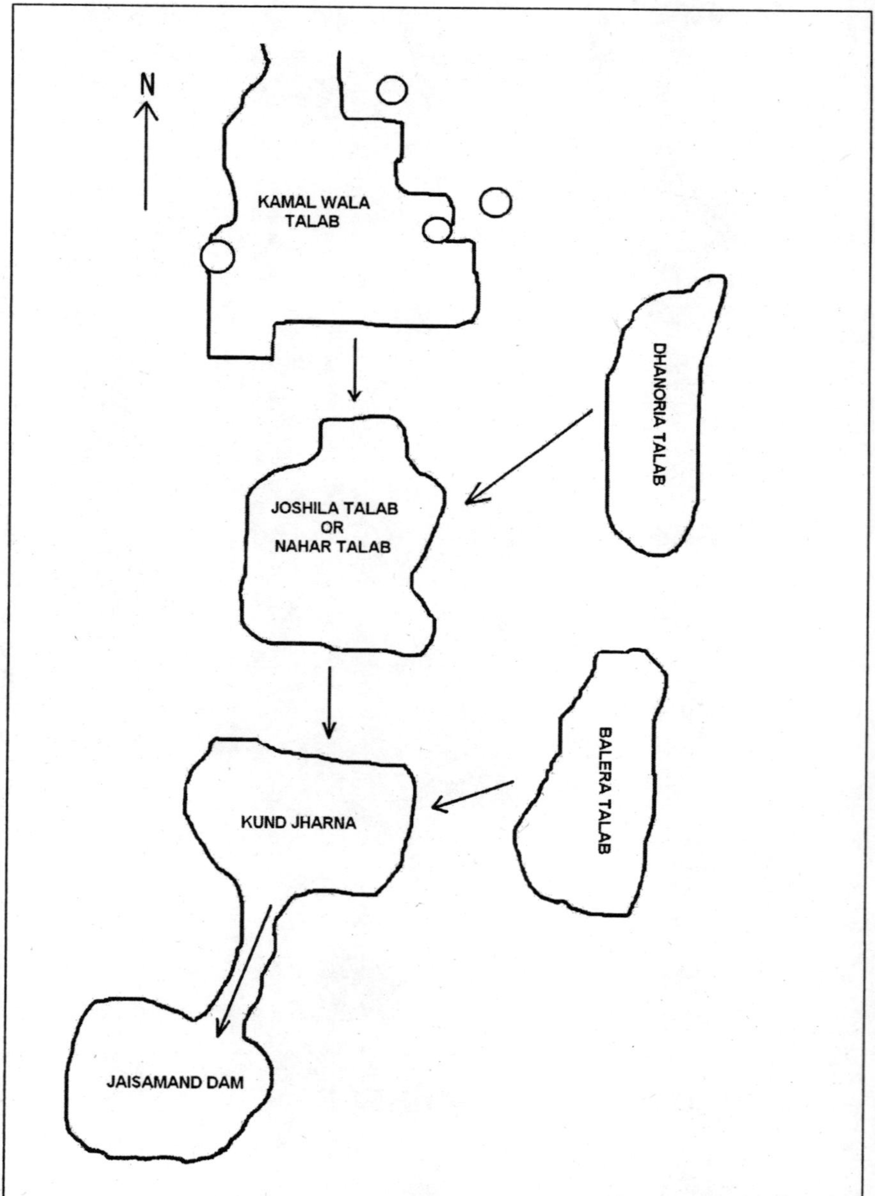

Figure 9.4: Chart of Connecting Tanks of Kanor

south-west of the village which receives water from the east. Its embankment wall runs east-west and gradually turns towards south. The length of the wall is about 750 m. while its breadth is 525 m. and its thickness is 8 m. The waterbody is located on the corner of a vast agricultural field. It is said to have been excavated by the peasants of the area. There is a strong possibility that the owners of the fields watered their crops with the help of the tank, the water being lifted with the help of the *dhenkli* and taken to individual field through channels (*nalis*).

The existence of such a large reservoir certainly raised the water-table to near the surface, which is evident from a step well used for watering agricultural fields.

The outlet is on the west and its surplus water first goes to Kareswar Mahadev tank and then falls into another anonymous tank. Thus the surplus water of numerous waterbodies[13] flows through *nallahs* and takes the form of the river Thaveria which merges into river Gomti, which falls into the Jai Samand. Ground plan of tank is prepared with its outlet (Figure 9.5).

Thus all the villages, namely, Bhindar, Kanor, Boheda, Mahuda and Bansi contain waterbodies whose surplus water similarly contributes to that of the Jai Samand.

The above tanks exhibit the efforts of the peasants in creating a network of reservoirs in their villages located far away from the capital.

Man Sagar

Man Sagar is located in Bansi, formerly a village of *pargana* Sadri.[14] It is situated 24° 20' N. and 74° 24' E, about 75 km south-east of Udaipur city. It was the headquarter of the *thikana* which was owned by the Shaktawat clan of the Sisodias. It was said to have been assigned as hereditary *patta* or *jagir* to Kesari Singh by Maharana Raj Singh.[15]

The tank named Man Sagar[16] is quite large and receives water from the hills located in the south-east. The retaining wall runs from south to north and is made of mud and boulders. There are

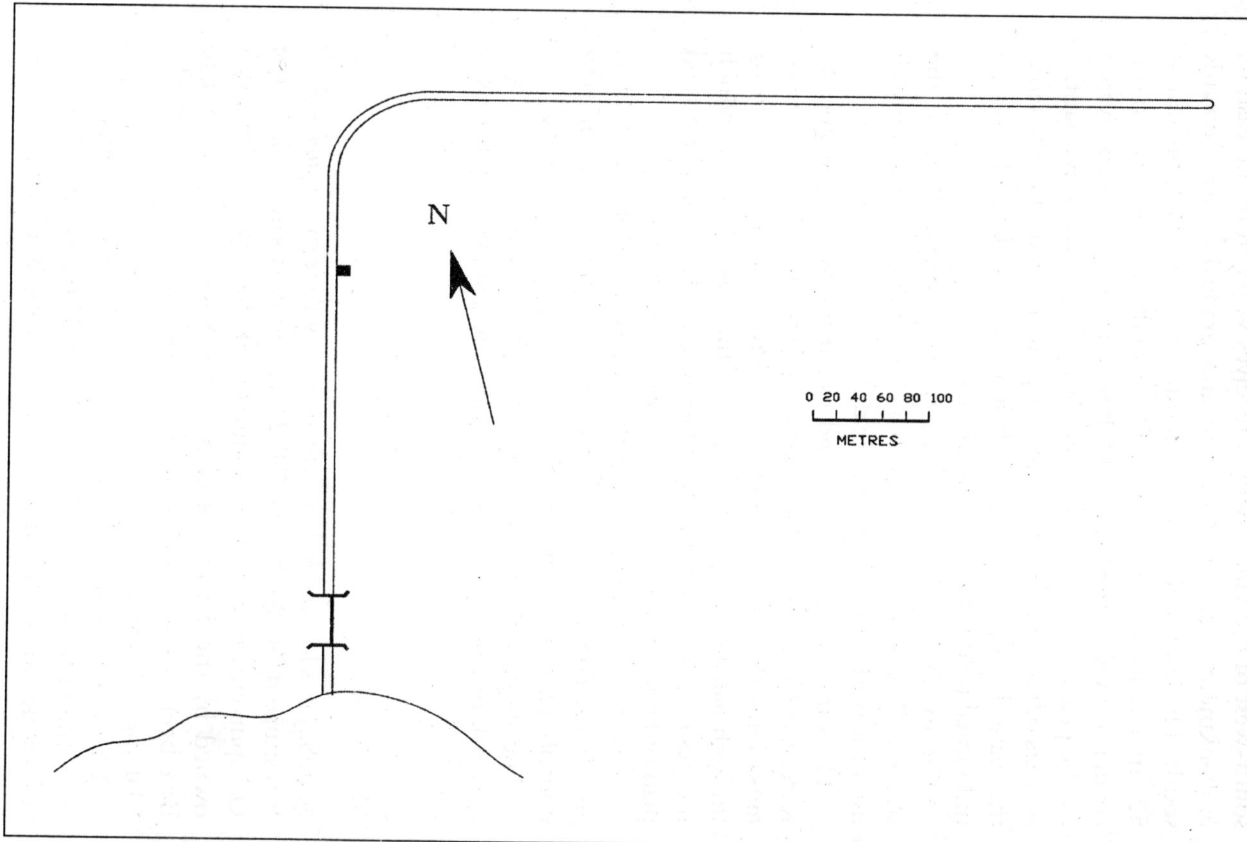

Figure 9.5: Ground Plan of *Maḥurda ka Talab*

six wells on the bank of the waterbody and one well is in the middle of the field.

Another well is located 200 m. away to the east of the tank, which is nowadays called the *Tikamji ka Kuwa* which obviously indicates private ownership. The agricultural field of the family is attached to it. The position of another well is in the middle of the field towards the western direction. It is approximately 100 m. away from the embankment wall.

Another water-monument known as the *Amrit Kund* is positioned 1 km. to the south of the tank. It appears that the overflow and the percolation of this waterbody feeds the *kund*. On the western side is a stone platform (*dabra*) with two horizontal slabs toward the side of the fields. This shows that this waterbody was meant for irrigation.

In addition to these, there are three step wells meant for irrigation in the area. This suggests that the groundwater level reached near the surface because of the existence of the Man Sagar tank. This reservoir fed many irrigation wells and step wells.

Local people say that the overflow of this tank contributes to the waters of the Jai Samand; this is confirmed by the depiction on the map (Figure 9.6). The water from the Mahurda and Bansi together form the river Thaveri which flows to the above-mentioned lake.

Moreover, the evidence contained in the inscriptions and literary compositions can be computed and put in Table 9.1.

Though Tables 9.1 and 9.2 do not contain entire data on the tanks built by the Sisodia rulers and the *thikanedars*, they are certainly an index of the rulers' interest in the construction of the waterbodies in their area of occupation. A large number of extant water structures demonstrates the partnership among the rulers, ruling elites and the people. This had changed the landscape of the region in medieval times.

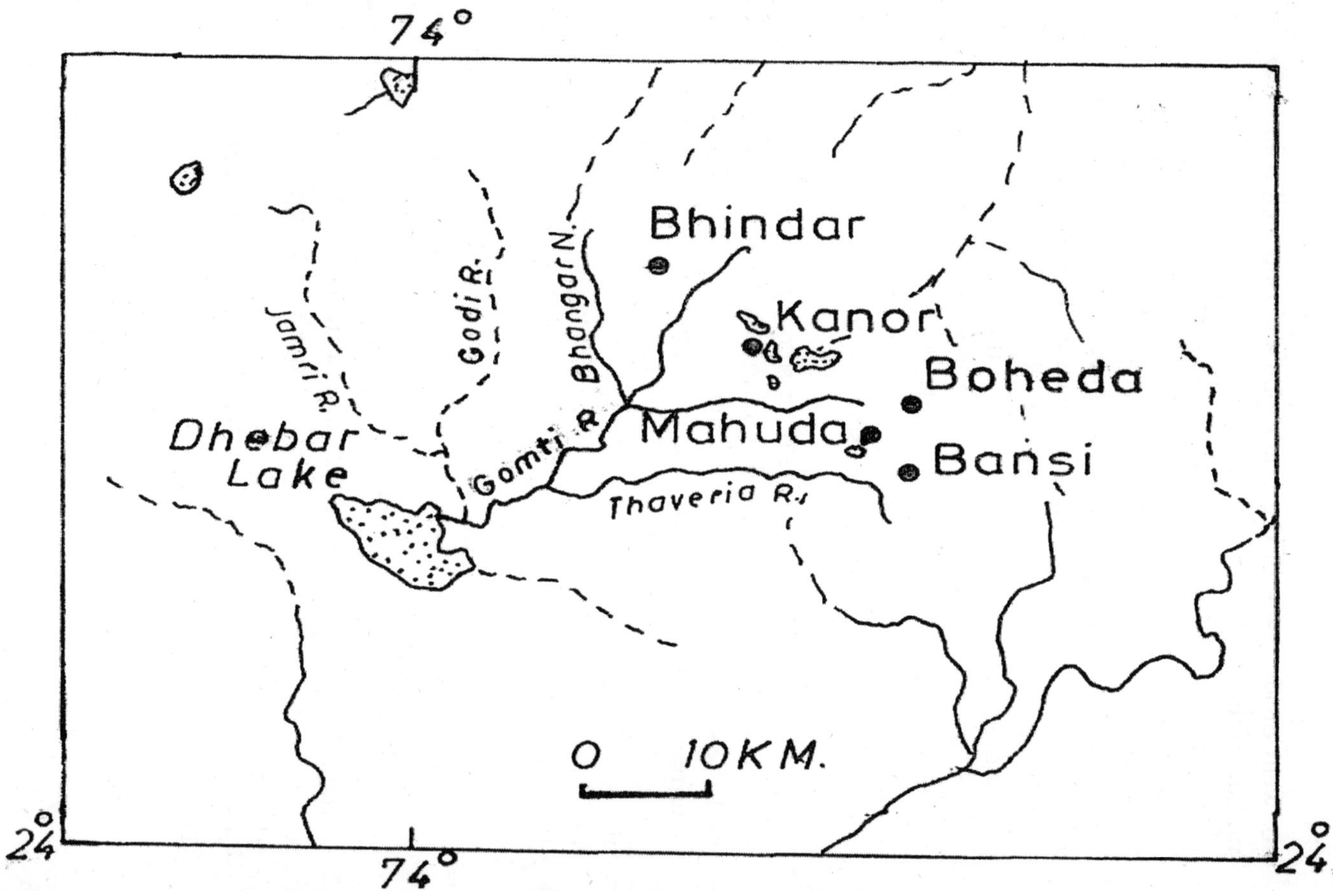

Figure 9.6: Jai Samand Feeder Tanks

TABLE 9.1: TANKS BUILTS BY THE RULERS OF MEWAR

Designation of the waterbody	Name of the ruler & their period of reign	Place	Period of Construction	Reference
Ranchhod	Shiladitya	Near Arnod	–	Ranchhod Bhatt, *Amarkavyam*, pp. 77–9
Bhalig (Bholela)	Bhogaditya	Near Eklingji	–	Ibid., p. 80; '*Sisod Vanshawali*' in *Majhamika*, no. 23, Udaipur, 1995, p. 22
Bappsar	Bappa	Guchha near village Ghasa		Ibid., p. 103
Hamirsar	Hamir	Kelwadpur (Kelwara)		Ibid., p. 134; '*Sisod Vanshawali*', op. cit., p. 43
Tanks (22 in numbers)	-do-	–	–	It is said that he built 22 tanks at different locations during his reign, cf. *Amarkavyam*, p. 140
Lakhesar	Lakha	Chittor	–	'*Sisod Vanshawali*', op. cit., p. 45
Lekhawali	-do-	Kelwara	–	Ibid.
Khetela	Kheta	Gogunda	–	*Gogunda ki Khyat*, ed. H.S. Bhati, p. 78, '*Sisod Vanshawali*', op. cit., p. 45
Baghela	Mokal	Nagda	–	Ibid., p. 150
Moga Nado	-do-	-do-	–	'*Sisod Vanshawali*', op. cit., p. 46
Tank	Bagh Singh (brother of Rana Mokal)	Madar	–	Ibid
Sar	Kumbha	Badhnor	–	*Amarkavyam*, op. cit., p. 162

Contd.

Tank	-do-	Kumbhalmer	–	Ibid., p. 166
Tanks	-do-	Shri Basantpur	–	In Abu Kumbha is said to have built seven tanks, ibid., p.168
Tank	-do-	Achaleswar	–	'Kritestambha' in Ojha, *Udaipur Rajya ka Itihas I*, p. 312
Tank	Rambai, sister of Kumbha	Kambhalgarh	–	Prashasti of Ramswami Temple, *shlok* 12; '*Sisod Vanshawali*', op. cit., p. 50
Ramsar	Raimal	Chittor	–	*Amarkavyam*, op. cit., p.181
Shankarsar	-do-	Kumbhalmer	–	-do-
Samayasankat	-do-	–	–	Bhavnagar Inscriptions, p.121 cited by Ojha, *Udaipur*, op. cit., p. 344
Jawar Talab	Maharana Amar Singh	Jawar	–	'*Sisod Vanshawali*', op. cit., p. 59
Rup Sagar Talab	Maharana Jagat Singh I	Ahar (Udaipur)		*Jagannath Rai Prashasti, shlok* 34
Jawar Talab	-do-	Kelwara	–	'*Sisod Vanshawali*', p. 61
Debari Ka Talab	Prince Raj Singh	Debari		*Vir Vinod*, II, p. 476
Surtan Sagar	Prince Surtan son of Maharana	Udaipur	–	Cf. Ishwar Singh Ranawat, op. cit., p. 103
Kalalya	-do-	-do-	–	-do-
Rangsar	Maharana Raj Singh	Udaipur	VS 1725/ AD 1668	Ojha, *Udaipur Rajya Ka Itihas*, II, p. 575
Tank	Kachla (Kachhola)	Not known	Mandalgarh	Brijmohan Jawaliya 'Mewar Rajya Men Mandalgarh Pargana Ke Gaon Ki Vigat' in *Aayam*, no. 1, 1994, Bhilwasa, pp. 20–44
Mandalgarh	Mandalgarh	Not known	Mandalgarh	Ibid.

TABLE 9.2: TANKS BUILT BY THE *THIKANEDARS*

Designation of the tank	Name of the *Thikanedar*	*Thikana*	Year of the Construction	Reference
Rajgarh ka Talab	Rawat Raj Singh of Begun	Begun	AD 1662	*Tawarikh Begun* (Shyamal Das Collection), cf. Ishwar Singh Ranawat
Talab	Not known	Qasba Ramgarh Bedla	N.A.	'Thikana Bedla Ka Itihas' in *Majjhamika*, no. 37, AD 2002, p. 4
Bakhat Sagar	N.A.	Do	N.A.	Do
Talab	N.A.	Kunwaliya (Bedla)	N.A.	Do
Talab	N.A.	Pandoli (Bedla)	N.A.	Do
Talab	N.A.	Kojuda (Bedla)	N.A.	Do
Talab	N.A.	Junda (Bedla)	N.A.	Do
Tanks	N.A.	(Bedla)	N.A.	Every village contains small tanks, ibid.

NOTES

1. Two irrigation tanks, the Sudarshana Lake by Chandragupta Maurya and another at Bhojpur, Malwa, constructed by Raja Bhoja in the eleventh century, show the antiquity of the reservoirs, cf. for the inscription of the former lake, see R.N. Mehta, *Journal of Oriental Institute*, XIII [1 and 2], pp. 20–38; for the Bhojpur reservoir see, D.D. Kosambi, *Introduction to the Study of Indian History*, rev. 2nd edn., 1975, p. 302.
2. Erskine, op. cit., pp. 46–7.
3. R.C. Sharma highlights the role of water in the settlement of the region. He writes that the wells were significant in deciding the location of most of the villages in the western areas while in the southern part, the tanks or ponds controlled the site of the village. (*Settlement Geography of the Indian Desert*, New Delhi: Kumar Brothers, 1972, p. 22.)
4. R. Thomas Rosin, 'Locality and Frontier Securing Livelihood in the Aravalli Zone of Central Rajasthan', in Karine Schomer, Joan Erdman, Deryck O. Lodrick and Lloyd I. Rudolph (eds.), *The Idea of Rajasthan*, Delhi: Manohar, 2001, pp. 35–6.
5. *EI*, XX, pp. 122–5 and XXXIV, pp. 173–6 cited by B.D. Chattopadhyaya, 'Irrigation in Early Medieval Rajasthan', in *The Making of Early Medieval India*, Delhi: Oxford University Press, 1994, p. 42.
6. For the Raj Samand and Jai Samand the terms Sagararaj and Samand are employed respectively, cf. Yati Ṣhri Khetal, 'Udaipur-Ki-Gajal', in *Nagar Varnatmak Hindi Padya Sangrah*, VS 1757/AD 1700, op. cit., p. 20. For the use of the term *sarovar*, cf. Man, *Raj Vilas*, op. cit., p. 99.
7. In Sanskrit the term for big dam is used *setu*, cf. Ranchhod Bhatt, *Amarkavyam*, op. cit., p. 83.
8. In addition to the *talab* or *talav*, the term *talap* was also employed, cf. Shri Khetal, 'Udaipur Gajal', op. cit., p. 21.
9. The Sanskrit terms *tadag* and *sar* are synonymous with tank, cf., *Amarkavyam*, op. cit., pp. 103 and 181.
10. The *pargana* was assigned in *patta* to Sarangdev, the son of Maha Singh, in VS 1768/AD 1711 (see copy of the original *patta* in Jamnesh Kumar Ojha, 'Kanor Jagir Ka Patta', in *Sodh-Sadhna*, no. 4, Sitamau:

Natnagar Shodh Sansthan, 1983, pp. 45–50; for geographical location, see Erskine, op. cit., p. 113.

11. The ownership of the well located to the north near the *chhatris* vests with a person named Narain Lal Pujari. His land is adjoining the well from which he watered his fields. He told us that formerly this well was jointly owned by him and his two brothers. Later on, his two brothers sold their rights to him and therefore now it is his property. Thus the *pujari* exercises exclusive right over the water monument.
12. The full name of the *pargana* is Sadri Kudal (cf., *Maharana Raj Singh Pargana Bahi* 2, ed. Hukum Singh Bhati, Udaipur: Himanshu Publications, 1995, p. 145. Presently its position was reduced to a tehsil of the district Chittorgarh.
13. Though the people interviewed said there used to be seven tanks in the past we did not notice any remains there.
14. *Maharana Raj Singh Pargana Bahi*, op. cit., p. 14. The estimated revenue of the village was Rs. 1,501.
15. Erskine, op. cit., p. 92.
16. The nomenclature of the waterbody is mentioned in an inscription of VS 1894/AD 1837. It appears that the tank came into existence after the permanent assignment of Bansi to Kesri Singh Shakhawat by Maharana Raj Singh (1652–80). Just after that this site was selected for this tank. Thereafter, it was repaired whenever the need arose.

CHAPTER 10

Foundation of the Capital City and Water Harvesting Devices

The medieval city of Udaipur, founded in the sixteenth century for its geographical setting and natural and man-made elegance, has always fascinated people and has earned the appellation of 'City of Lakes'. The city's beauty received rightful appreciation from James Tod. He remarked that the valley of Udaipur is 'the most diversified and most romantic spot on the continent of India'.[1] Other writings too echo similar views. Such laudations comprehend the contribution of both nature and man in making Udaipur 'romantic'. Before 1550, the area despite its exotic natural beauty did not get adequate attention from anyone. It is believed that this valley acquired prominence in March 1559 when Maharana Udai Singh, the chief of the Sisodia clan and the reigning monarch of Chittor, so fascinated by its natural beauty, decided to develop it as his capital (Figure 10.1).

At the outset, three aspects need attention: first, the foundation of the new city of Udaipur and its layout; and second, the construction of the Udai Sagar Dam which was the lifeline and part and parcel of the city and last, the organization of the water supply system which was an essential part of the urban settlement.

THE NEW CAPITAL CITY AND THE ERECTION OF WATER HARVESTING STRUCTURES

In the year 1537, Maharana Udai Singh successfully dealt with the palace intrigues of the imposter Vikramaditya and ascended the throne of Chittor. Later, on a visit to Eklingji (the deity of the Mewar house) he went to Ahar for hunting. The topography and

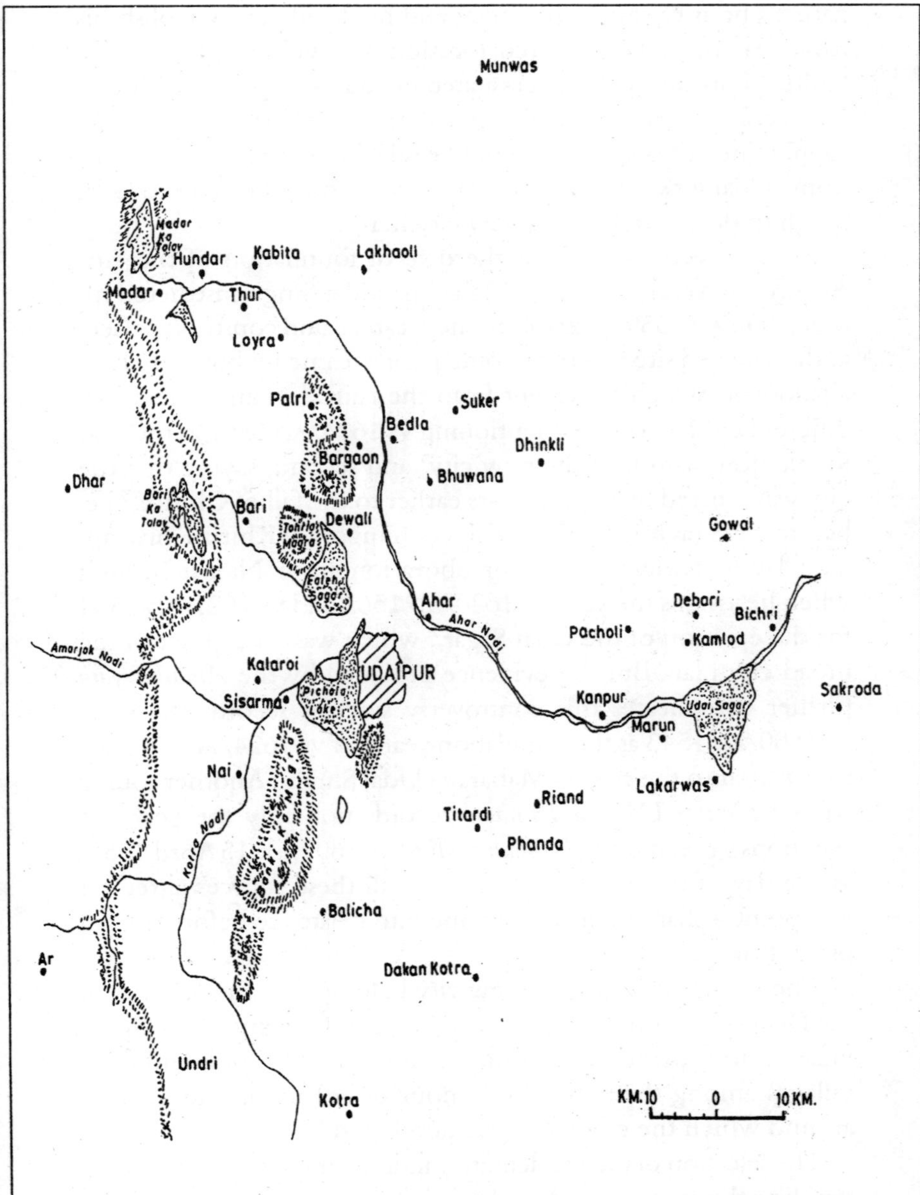

Figure 10.1: Udaipur and its Environs

nature's bounty caught his fancy, and he decided to establish his royal seat here. The isolated location and vulnerability of the Chittor Fort on a ridge necessitated the quick implementation of his decision: Chittor's weakness was that once besieged and the supply line cut off, Chittor could easily fall to enemies. So apart from Udaipur's natural beauty, military consideration forced Udai Singh to decide to found a new city here.

A controversy surrounds the date of foundation of the city. According to Shyamal Das, Maharana Udai Singh instructed his officials in AD 1559 to establish a new capital and construct palaces at the selected site.[2] Initially, some palaces came up but were later abandoned which is evident from the ruins.[3] Nainsi offers two different evidences, one mentioning VS 1624/AD 1567 as the year for the foundation of the new city[4] and second, saying that the city was erected five or ten years earlier to the fall of Chittor,[5] i.e. before 1568 (as told to him by his contemporary Khiria Khiwraj). The latter evidence gets corroboration from Nainsi himself when he assigns the year AD 1620/AD 1563 and VS 1621/AD 1564 for the erection of the Udai Sagar,[6] which was an adjunct of the broad city plan. But the evidence provided by the *Amarkavyam* further complicates this controversy since it records the year VS 1600/AD 1543 as the foundation year and VS 1624/AD 1567 for the entry into the city by Maharana Udai Singh.[7] Another source named *Chittor Udaipur Patnama* records not only the year but mentions a specific day, *Baisakh Sudi* 3, VS 1609, i.e. 15 April 1553, as the day of its founding.[8] However, all these sources agree on one point—that the new city came into existence before the fall of Chittor.

The site selected for the new city belonged to the *bhumias* of the Devra clan, comprising 52 villages collectively known as Girwa villages, and surrounded by mountains.[9] The Pichholi and Palri villages among them got the honour of becoming the nucleus around which the city had to be developed.[10]

The selection of the site definitely indicate the wisdom of understanding the strategic and aesthetic importance of sites for establishing a city or building. Here, the Maharana himself selected the location for the new city. After a vigorous scrutiny of the topography of the area, he prepared a plan which he put before the nobles of the state.[11] He pointed out that the site was close to a tank of enormous magnitude known as Pichhola.[12] The palaces were to be erected along its bank and the town on its east and north. The fort was to be raised on the hill top in the south for protection and safety of both people and royalty.

Further, the Maharana visualized the location of the outer gates. At the extremity of a hill, at a distance of 12 to 14 km. from the Pichhola, the ground was to be prepared after levelling the mountain and then a gate (*darwaza*) be raised to serve as a fort. Again, another fort was to be built-up in the north at the end of the *Dungar ki Nal* which was named after the village.[13]

Maharana Udai Singh also put out his plan for the construction of a tank. On the eastern direction, a deep gorge existed around which there was a cluster of villages owned by the Devra Rajputs. They were to be settled in other places. A big tank was to be created after erecting an embankment wall between two hills.[14]

The above plan was presented before the nobles by the Maharana, which demonstrated his broad vision about a future city. The plan encompassed three most important things: one, strategic and military viability of the chosen site; two, interrupted water supply to the city dwellers; and the third, a hinterland to supply food to all.

It shows that the Maharana had the acumen of a civil and military engineer. The location selected by him was surrounded by high hills as natural defence, making the future city impregnable for enemy forces. It also meant less burden on the state exchequer. The construction of the fort on the hill-top was aimed at strengthening the defence, providing protection to the royal palaces and to the common folk. Three gates were to be constructed in different directions to ensure foolproof defence in case of an invasion.

The selection of the Pichhola Lake as a water source for the new city is an indicator of the vision and wisdom of the Sisodia ruler. Water from the two hills, Machhla and Sisarama, flows into the tank. The populace of the city solely depended on the tank for their water needs. The lake's surplus water overflows into the river Berach,[15] showing its vastness. The third aspect conceived by the Maharana was a dam which was to be built between the two hills. His intention was to create a tank for irrigating the surrounding areas. This way, he wanted to have a hinterland for regular supply of food for the city dwellers.[16]

After unfolding the 'blueprint' of his plan, instructions were issued to civil engineers to translate the plan into reality. It is interesting to know how the plan was carried out. The engineers initiated work by first constructing the royal palaces. Since the first right to water was the royal prerogative, the engineers erected royal palaces on the embankment (*pal*) of the Pichhola (Figure 10.2). On the west, the city was founded.[17] Nainsi records the size of the city after about a century of its foundation: it was about 8 km. in circumference.[18] While another contemporary writer Kavi Maan who, completed his work in and around 1670 records the bounds of the city as 48 km.[19] His statement is also confirmed by later poets.[20] This was an appreciable expansion of a town in hundred years, showing a steady increase of its population.

On the location suggested by Udai Singh, an enormous tank was erected by constructing a masonry wall. The *Ain-i Akbari* records its circumference as 64 km.[21] while Nainsi gives 80 km.[22] Anyway, its huge size is confirmed by both the sources. The size of the embankment was appreciably big. Its wall was 305 m. in length and 152.5 m. in breadth. Its height came to 42.7 m.[23]

The plan of constructing canals appears to have been carried out systematically as envisaged by Udai Singh. A long and deep canal was cut out of the hill for irrigating a vast tract of agricultural land.[24] Abul Fazl notices cultivation of wheat in a large tract.[25] Similarly, a canal (*mori*) was dug out of the Pichhola Lake which watered the surrounding area. It brought enormous *hasil* (revenue) to the state treasury.[26] The canal-irrigated area proved a strong lifeline for the city. It must have certainly provided cash crops as well as food grains to the city people. The fertility of this

Figure 10.2: Photo of Pichhola Lake and Palaces

area received appreciation from James Tod. He was overwhelmed by this valley and called it *doab*, watered by the Berach and Banas rivers. He further mentions that 'there is not a more fertile tract in India than this, which would alone defray the expenses of the court if its resources were properly husbanded'.[27] This observation confirms the wisdom and topographical knowledge of those who had chosen this site for the new city.

Buildings, Gardens, Water Harvesting and Management

The founding ceremony heralded the beginning of constructional activity in the new city. In the next four years between 1552 and 1556, the walls of the Pichhola tank, fort and *topkhana*, palaces and temples of Eklingji were completed.[28]

Interestingly, the wives of the Maharana also contributed a lot to the new city by constructing several structures of public utility. They built *sarais* or *dharmasalas* along different routes. The Songiri wife settled a village about 3.2 km. away from the Debari gate, and a *sarai* with a step well (*baori*) was erected there in VS 1611/ AD 1654-5. This came to be known as *Songiriji ki Sarai*. It is also called *Teliyon ki Sarai*[29] because the local settlement of the telis was there. Similarly, Sahaj Kanwar, the *Rani* of the Solanki clan, got constructed a *sarai/dharmasala* with a step well in VS 1612/ AD 1555-6 which was named after her.[30] Their examples were followed by the Rana's wives from the Jhala, Chauhan, Khinchi and Bhati clans.[31] All the *sarais/dharmasalas* built by them bore their names to indicate their contribution in the construction of these structures.[32] Other wives also took active part in the construction activity in the city.

The construction of the water harvesting structures and other buildings such as wells, step wells and *sarais/dharmasalas* by the wives of the Maharana was a significant contribution in the development of the city. Their location along different routes offered facilities to travellers coming from diverse places. This facilitated caravan trade also and attracted a large number of merchants and traders to the city. The continuous development of the city attracted a large number of *mahajans* and craftsmen to settle in the city. The population of *mahajans* and artisans rose to about 50,000 in VS 1624/AD 1567.[33] This is an indication of the peace that prevailed in the city, leading to the fast growth and development of its overall population and its economic activity.

The gardens developed under royal patronage occupy a prominent place in the planning of the city. Their placement at different locations display royal attempt at beautification and to use them as symbols of royalty. After the establishment of the new capital city Udaipur in the 1550s, it remained the centre of power till independence and enjoyed utmost glory. During this period, successive rulers, their wives and their noblemen created innumerable gardens in the city and elsewhere in Mewar. Though Maharana Udai Singh laid out his capital town with an enormous lake, he could not get an opportunity to create gardens. This work was accomplished by succeeding rulers between 1620 and 1730. The pioneer in the developing gardens was Maharana Karan Singh (1620–8) whose reign was most peaceful and prosperous. He laid the first garden outside the Udai Pol, known as the Krishna Vihar.[34] It was full of various kinds of trees and flowers which were watered by numerous fountains. Similarly, he created another garden named Champabagh which too was adorned with fountains. In the middle of it was built a pleasure building.[35] Though we do not get much evidence about these pleasure sites, we can say that these gardens were developed on the Mughal pattern because Karan Singh was the first Sisodia prince who came into contact with the Mughals and got acquainted with the Mughal way of life.

During his princehood Raj Singh laid a marvellous garden and magnificent buildings at Udaipur (on the east of the Sajjan Niwas), known as the *Sarvaritu Vilas* (pleasure garden for all seasons).[36] It is in an enclosed compound with entrance in the south and numerous sculptured buildings of magnificence. The provision of continuous water supply was developed by the constructions of a step well.[37] The water was guided to different subsections of the garden. On all the four sides numerous fountains were installed[38] to make the surroundings cool and pleasant. The plantation was rich with various fruit-bearing plants including oranges, mangoes, grapes, pomegranates, dates, pistachios, almonds and walnuts.[39] Some fruit plants appear to have been imported from other regions.

In addition, innumerable gardens were noticed by Nainsi[40] and other authorities in and around the city.[41] Some of them, though small in size and were styled as *Wadi* or *Badi* (small, enclosed garden), for example, the *Neelkanth Wadi*, *Nagina Wadi*, *Setha ki Wadi*, etc. These appear to have been laid by private individuals, indicating the participation of people in the creation of gardens.

There is an interrelationship between climate and gardens.[42] This gets support from the contemporary poets who noticed clean environment near the gardens. Varieties of flowers attracted birds and spread fragrance. The singing of different kinds of birds provided natural melody.[43] The areas around gardens became extremely pleasant for a stroll. It appears that the city-planners might have given serious thought to the environment and consequently, laid gardens within enclosed walls with beautiful buildings inside.

RISE IN POPULATION AND THE DEVELOPMENT OF URBAN WATER SUPPLY SYSTEM

The fast expansion of the newly-founded capital town in territorial terms as well as the enlargement of population in the next hundred years shows its attraction and charm, both administrative and environmental, which lured heterogeneous people to make this place their permanent abode. The establishment of the royal seat and the maintenance of law and order on trade routes naturally gave boost to commercial activities, consequently there was a inflow of merchants and artisan groups from different parts of the state and outside. The continuous flow of people enlarged the city both in terms of its size and population.

The city reached 48 km. in circumference in the next hundred years, i.e. the 1660s.[44] We may assume corresponding enlargement of population in this span of time. This augmentation is confirmed by the data on the house-count furnished by Nainsi. The total house-count was 20,060 in the 1660s,[45] which enables us to obtain the total number of persons by multiplying with the conventional demographer's rate of 4.5 persons per house. Based on this we would get a total population of 90,270 for the city. This is probably on the lower side, since each house must have contained more than 4.5 persons.

Moreover, the houses counted might not have included the houses of outcastes (like Bhangis and Chamars). If we assume that a raising of the population would be permissible, say by counting 6 persons per house (to cover under-enumeration of houses as well), we may put the estimated population *c.* 1660 at 120,360. This would rank Udaipur among the medium-size towns of the Mughal Empire.[46]

The development of a water supply system to the city residents was an essential part of urban planning without which human settlement on any scale would not have been possible, especially in a hilly region. As stated earlier, the original planner Udai Singh selected the site for his new capital on the bank of the Pichhola Lake on whose water the city population depended for their domestic needs.[47] The Maharana extended the embankment wall in the north up to the Machhla hill to enhance the lake's capacity to retain water. Its width too was increased to face the force of water. This wall is known as the *Bari Pal.*[48] In addition, an enormous tank, Udai Sagar was erected for irrigating the surrounding areas for enduring provision of food for the urban residents.

Many step wells were built by the wives of the rulers between 1555 and 1557 for providing drinking water to the residents of the area.[49] The work of the water harvesting systems in the city got a setback later after the demise of Udai Singh in February 1572 and the battle of Haldighati in 1576 when Mewar was occupied by the Mughals. The work of water management was restarted when the possession of Mewar came once again under the Sisodias after the treaty with the Mughals in 1615. Maharana Amar Singh is said to have extended the length of the Pichhola embankment wall towards the Sisarama hills, probably to further enhance the storing capacity of the lake.[50] After that more water was made available to the residents. The city got numerous water-bodies during the time of Maharana Raj Singh. He built the famous *Sarvaritu Vilas Baori* in the vicinity of the royal palaces.[51] He appears to have taken interest in water harvesting projects for providing sufficient drinking water to the residents of Udaipur and for irrigating the surrounding areas. He built *Rang Sagar* in the city itself and the *Bari ka Talab*[52] 8 km. north-west of the city. The overflow of the latter tank reached near the city, raising the water level near to the surface of that area.

Maharana Jai Singh took interest in the water supply system for the urban people. On his instructions two tanks at Dewali and Thur[53] were dug out to harvest rain water coming from the north-western hills. On the same site now stands the Fatah Sagar. Numerous other tanks such as Rup Kund, Swarup Sagar, Dudh Talai and Mulla Talai used to serve the local populace.[54] Besides tanks, a large number of step wells were erected by the members of the royal family, nobles and other rich people.[55] Most of the wells were dug out on the banks of rivers and streams.

The existence of a large number of water structures shows the interest of the Sisodia rulers in the development of urban settlement. Their interest is evident from their projects on water harvesting and conservation to provide drinking water to the city residents. Their wives too took keen interest in constructing waterbodies in and around the city. Besides, the rulers also developed waterbodies for the peripheral areas for irrigating the surrounding areas to grow crops to feed the city dwellers (Figure 10.3).

Their concern for environmental issues is seen in the construction of a number of gardens which played a positive role in providing a healthy and clean atmosphere, leading to the expansion of the city as well as population. Their interest was well received in commercial circles and consequently trade and commerce in the city accelerated and merchant groups came to settle there.

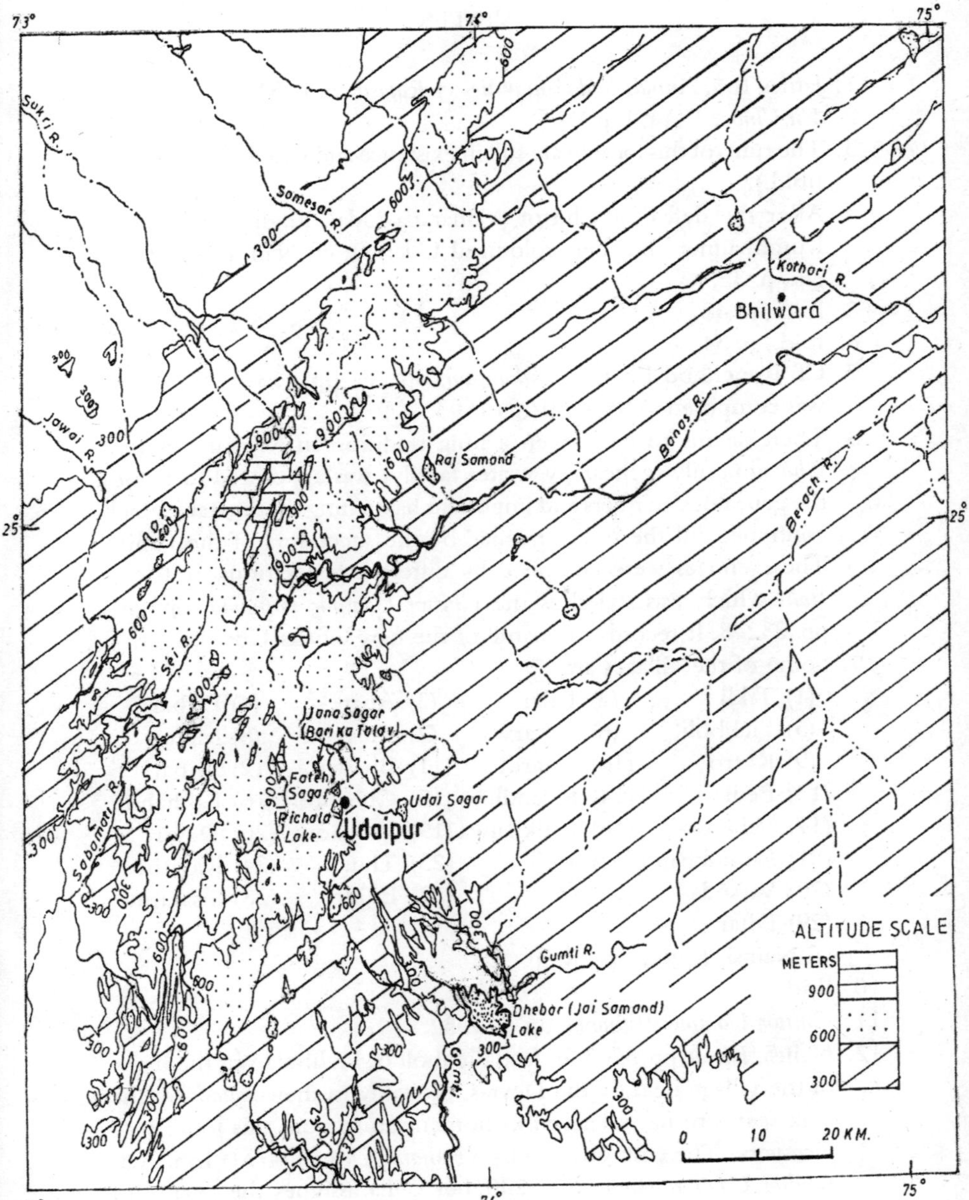

Figure 10.3: Waterbodies in and Around Udaipur City

NOTES

1. James Tod, *Annals and Antiquities of Rajasthan*, p. 519.
2. *Vir Vinod*, II, pt. I, p. 72.
3. The ruins of this locality are still in existence and styled as Motimahal (ibid.).
4. After the defeat at Chittor, Maharana Udai Singh first went to Kumbhalmer and then colonized Udaipur, cf. Nainsi, *Khyat*, op. cit., p. 32.
5. Ibid., p. 48.
6. Ibid., p. 34.
7. Cf. Ranchhod Bhatt, *Amarkavyam*, op. cit., pp. 213–15. The epic was composed in VS 1750/AD 1693.
8. There seems to have been a long tradition in Mewar where a *Charan* family (genealogy writer family) was appointed to record the genealogy of rulers and important happenings in the state which continued till the independence by the successors of the family. These entries were compiled by different generations in a long *Bahi* which was styled as the *Chittor Udaipur Patnama*, op. cit., pp. 232–8. It records the name of the new city as Udai Nagar.
9. Some of the villages are:

(1) Tikli	(2) Palri	(3) Ahar	(4) Dabari
(5) Pichholi	(6) Lakarwa	(7) Kalarwa	(8) Matun
(9) Kotro	(10) Titardi	(11) Bhawno	(12) Anbari
(13) Bedlo	(14) Ruandh	(15) Chhaproli	(16) Lakhaholi
(17) Veharwo	(18) Chikhalwa	(19) Vadgaon	(20) Devri
(21) Mundkhasol	(22) Vadi	(23) Thur	(24) Kavitho
(25) Varsoda	(26) Nai	(27) Bujdo	(28) Sisaramo
(29) Dhar			

   cf. Nainsi, *Khyat*, op. cit., p. 32.
10. Ibid.
11. *Chittor Udaipur Patnama*, pp. 229–32.
12. Nainsi, *Khyat*, pp. 32–3. In fact, Pichholi or Pichhola was the name of the village held by the Devras, probably in their *watan*. So the tank seems to have derived its nomenclature from this locality or vice versa. The tank was dug by a *banjara* in VS 1444/AD 1387. Cf. *Chittor Udaipur Patnama*, p. 230, but Ojha assignes the period of construction between 1382 and 1397, at the time of Maharana Lakha, cf. *Udaipur*, op. cit., p. 7.

13. *Chittor Udaipur Patnama,* pp. 231–2.
14. Nainsi, *Khyat*, pp. 21–34; *Chittor Udaipur Patnama*, p. 231.
15. Nainsi, op. cit., p. 33.
16. Ibid., p. 21.
17. Ibid., p. 32.
18. Ibid.
19. Kavi Maan, *Raj Vilas*, op. cit., p. 32.
20. Jaswant Sagar, 'Udaipur Varnan Chand' in *Maru Bharti*, January, 1959, no. 4, p. 40.
21. *Ain-i-Akbari*, p. 278.
22. Nainsi, op. cit., p. 35.
23. Ibid., p. 34.
24. 30.5 m. deep and 3.66 m. in width, *nallah* is cut out of the hill, cf. ibid., p. 34.
25. *Ain*, II, p. 278.
26. Nainsi, op. cit., p. 33.
27. Tod, op. cit., p. 619.
28. The cost of these constructions totals up to Rs. 36,91,917, *Chittor Udaipur Patnama*, pp. 238–40.
29. Ibid., pp. 238–9.
30. Ibid., p. 239.
31. Names of the wives of Udai Singh are Jhali Birabai, Chauhan Banras Kanwar, Khinchi Akhe Kanwar and Bai Ajan Kanwar (op. cit., pp. 239–40).
32. Ibid.
33. Nainsi records the number of the *mahajans* and craftsmen which were 2,000 and 9,000 respectively. Their population is obtained by multiplying by 4.5 persons per house, cf. *Khyat*, I, pp. 33–4.
34. Vyas, op. cit., p. 234.
35. Ibid.
36. Maan, op. cit., p. 57; *Rajprashasti*, op. cit., pp. 58–9.
37. Ibid.
38. Ibid., pp. 57–8.
39. Ibid.
40. Nainsi, op. cit., p. 33.
41. Khetal, 'Udaipur Gajal', op. cit., p. 21; Besides, a city map of Udaipur prepared in the eighteenth century gives the names of certain *badi* (small gardens) cf. G.M. Bahura and Chandramani Singh (eds.), *Catalogue of Historical Documents in Kapad Dwara, Jaipur*, Jaipur: n.d. II, pp. 95–6.

42. Maan, op. cit., pp. 57–8; 'Udaipur Gajal', p. 18.
43. Ibid.
44. Maan, op. cit., p. 32; 'Udaipur Varnan Chhand', op. cit., p. 40.
45. At one place Nainsi gives the total number of houses as 20,000, based on estimation while the total of the profession and caste-wise break up comes to 20,060, cf. *Khyat*, op. cit., pp. 33–4.
46. No doubt that the new capital Udaipur acquired the status of medium-size towns of the Mughal Empire but during the British period it lost some of its population. The city could not gain the same level of population till 1951. A comparison with modern census counts of the population of the city would be interesting:

Year	Source	Population
1660	*Khyat*	120,360
1881	Census	38,214
1891	,,	46,693
1901	,,	45,976
1911	,,	35,116
1921	,,	36,676
1931	,,	45,922
1941	,,	59,648
1951	,,	89,621

It becomes clear from the above table that the population of Udaipur was very high around 1660 compared to later years. In fact, there was a spectacular decline in it. The population did not cross even the 50,000 mark until after 1931, and was still below 100,000 in 1951. This may have special interest for those looking into the question of de-urbanization in the nineteenth century.
47. Nainsi, *Khyat*, I, p. 33. A *ghazal* records that the ladies took water from the *ghats* of the lake. Cf. '*Udaipur Ki Gajal*', op. cit., p. 15.
48. *Chittor–Udaipur Patnama*, p. 239. According to another source the extension of the embankment wall was carried out by Rana Amar Singh, cf. 'Mewar Rawal Ranaji-ri-bat', *Majjhamika*, pt. 17, Udaipur, 1993, p. 69.
49. Ibid., pp. 238–40.
50. See note 48.
51. Maan, op. cit., '*Udaipur Ki Gajal*', op. cit.

52. The Rang Sagar was constructed during the time of Raj Singh and its completion ceremony was performed by Prince Jai Singh, cf. *Rajprashasti*, pp. 88–9.
53. See *Maharana Raj Singh Pargana Bahi*, op. cit., pp. 1–2.
54. Vyas, op. cit., p. 166.
55. *Vir Vinod*, I, pp. 150–60. It furnishes a list of the *baoris* and tanks in Udaipur.

Conclusion

The objective of this study has been to trace and document the efforts of the Sisodia rulers in the area of water harvesting and conservation of rain and sub-soil water to be used for drinking as well as for irrigation. Numerous extant waterbodies such as *baoris*, tanks, wells and dams stand witness to the rulers' almost passionate endeavour in the field of water harvesting and conservation.

Besides embellishing the capital and other royal seats with a variety of hydro-structures and related monuments they also carried out water harvesting projects in the area under their control. It is true that their efforts were based on two objectives: first, to achieve the mythological status of Bhagirath (who is said to have brought the Ganga on the earth); and second, the more earthly utilitarian purpose, that is, drinking and irrigation. The latter can also be linked with the economic necessity of the state.

Moreover, the contribution of the *jagirdars* or *thikanedars* too was no less noteworthy. They undertook the projects of water harvesting and conservation in their respective *jagirs*. The *thikana* documents show the excavation of wells, step wells and tanks in abundance by them. The case of *thikana* Bedla, where a large number of wells, step wells and tanks were constructed by the *thikanedars* and their wives, stands as testimony to this. In addition, other agencies such as peasants, caste panchayats and traders and the *banjaras* were equally engaged in the excavation of wells, tanks and the step wells. A large number of wells in rural areas stand as witness to the people's or peasants' participation in this sector.

Through the construction of various waterbodies the indigenous hydraulic technology developed stage by stage. As experience in the construction field expanded, the water structures too got more refined. A dominant feature of the medieval Sisodia state was the construction of dams at regular intervals.

All the embankments built between AD 750 and 1700 could be characterized as gravity dams. The main characteristics of this kind of structure is a heavy body which could in real terms face the force of water. The hydraulic engineers appear to have put weight on the body after careful calculation of the thrust of the water which in turn depended on the quantum of water passing through the point where the embankment wall was erected. The study of extant dams demonstrates that the hydraulic engineers had developed expertise to erect an embankment best suited to the available site. It is evident that the size of the embankment increased as the experience of the engineers enlarged. One of the early dams, the Indra Sarovar, is small in size and light in weight while the Jai Samand is the largest in many respects. The latter is an index and symbol of refinement in the hydraulic technology. This kind of development in hydraulic technology was achieved through accumulated experience in building construction.

Three distinct technical advances were made in the long term: one, the dressed stone replaces the traditional rubble and laterite material, particularly for the construction of the dams; two, the stones were laid with precision which is evident from the mathematical calculation made by the mason on a stone slab of the body of the Raj Samand; and the third, the construction of stone-sluices at suitable places to regulate the supply of water from the lake.

In addition to dam construction, the hydro-engineers appear to have acquired expertise in exploiting sub-soil water for irrigation. This is evident from the wells excavated on the banks of rivers and tanks where water percolation raises the ground-water level. This kind of well, styled as the *seja kuwa* (shallow well), shows the people's knowledge in the field of hydrology.

Innumerable extant tanks were excavated through many agencies such as the state, *jagirdars*, village community or individuals, suggesting the engagement of expert hands. This is evident from the construction of strong *bunds* around the tanks. In some places rows of trees were planted on the tank *bund* evidently with the view to protect it from soil erosion. In addition, the task of connecting tanks located in the vicinity was accomplished with remarkable imagination and skill. Not only this, the ultimate overflow of all tanks of Kanor merged into the Jai Samand. Thus the surplus water of many tanks of the state was linked with one or other bigger lake. This shows the highest degree of technical skill of the hydro-engineers.

The interest of the Sisodia rulers in the organization of the water harvesting system for the urban people was equally noteworthy. Sufficient number of waterbodies was built to meet the growing demand of the enlarging population. The engineers of Maharana Raj Singh and Jai Singh endeavoured to link the waterbodies with each other to enlarge the volume of water available for the urban dwellers. Moreover, the study of water harvesting, conservation and irrigation of medieval Mewar shows a process of development in the field of technology.

APPENDIX

Units of Distance and Measurement Occur in the Medieval Literature of Mewar

There are numerous terms employed in medieval times to indicate distance and measurement, etc., which have now become almost obscure. Therefore, for the convenience of readers their equivalence is established in modern terminology.

It is pertinent to note that in Mewar *gaj* (Pers. *gaz*) was employed by masons which was equal to 2 ft known as the *silawati gaj* therefore, its equivalence in metres is established accordingly. Besides it, equivalence of *kos* is given in kilometres. Moreover, the size of the *bigha* is not known therefore it is left as it is.

1 *kos*	=	4 km.
1 *gaj* of 2 ft	=	0.61 m.
1 feet	=	0.305 m.
1 mile	=	1.6 km.
1 inch	=	25.4 mm.
1 sq. mile	=	2.56 sq. km.

Bibliography

RAJASTHANI, HINDI AND PERSIAN SOURCES (PUBLISHED AND UNPUBLISHED)

Babur, Zahiruddin Muhammed, *Baburnama*, tr. Annette Susanah Beveridge, I & II, New Delhi: Oriental Book Reprint Corporation, rpt., 1979.

Bahura, G.N. and Chandramani Singh (eds.), *Catalogue of Historical Documents in Kapad Dwara, Jaipur*, Jaipur: Maharaja Jaipur (n.d.).

Bhati, Hukum Singh (ed.), *Mewar ke Aitihasik Patte Parwane-Thikana Vijaygarh Sangrah AD 1605–1907*, Udaipur: Pratap Shodh Sansthan, Bhopal Nobles Sansthan, 1993.

——— (ed.), *Patta Bahi Pattedaran-ri-Vigat*, vol. I, Udaipur: Himanshu Publications, 1995.

Bhatt, Ranchhod, *Amarkavyam*, ed. S.K. Sharma and R.P. Bhatnagar, Udaipur: Sahitya Sansthan, Rajasthan Vidyapeeth, 1985.

———, *Rajprashashti Mahakavyam*, ed. M.L. Menaria, Udaipur: Sahitya Sansthan, Rajasthan Vidyapeeth, 1973.

Fazl, Abul, *Ain-i-Akbari*, vol. II, H.S. Jarret and 2nd edn., corrected and annotated by Jadunath Sarkar, New Delhi: Orient Books Reprint Corporation, 3rd edn., 1978.

Hem, 'Medpat Varnan Prashashti' and 'Mewar Desh Chhand', *Maru Bharti*, July, vol. II, pt. I, Pilani: Birla Education Trust, 1954.

Jai Singh Nrap Prashasti (Mss), Kolkata: Asiatic Society of Bengal.

Jawaliya, Mohan, 'Mewar Rajya Men Mandalgarh Pargana-ke-Gaon-ki-Vigat', in *Aayam*, ed. Devendra Kumawat, Bhilwara: Mewar Adhyayan Evam Vikas Sansthan, 1994.

Khetal, Shri, 'Udaipur ki Gajal', in *Shrinagar Varnatmak, Hindi-Padya Sangrah*, ed. Muni Kanti Sagar, Surat: Shri Jindutt Suri Gyan-Bhandar, 1948.

Maan, Kavi, *Raj Vilas*, ed. Motilal Menaria, Varanasi: Nagri Pracharani Sabha, VS 2015.

Maharana Raj Singh Pargana Bahi, vol. II, Udaipur: Himanshu Publications, 1995.

'Mewar Jagirdaran-ri-Vigat', in *Majjhamika*, no. 16, Udaipur: Pratap Shodh Sansthan, 1992.

'Mewar Rawal Ranaji-ri-Bat', *Majjhamika*, vol. 17, Udaipur: Pratap Shodh Sansthan, 1993.

Nainsi, Muhnot, *Khyat*, I, ed. B.P. Sakariya, Jodhpur: Rajasthan Oriental Research Institute, 1960.

———, *Marwar-ra-Pargana-ri-Vigat*, ed. N.S. Bhati, Jodhpur: RORI, 1968.

Ojha, J.K. (ed.), 'Kanor Thikane ke Patte Parwane', Udaipur: Pratap Shodh Sansthan, 1996.

Parwanas in Bilara Collection (Mss), Sitamau: Natnagar Shodh Sansthan.

Phutkar Khyat, Kaviraja Collection, *Granth* no. 109, Sitamau Natnagar Shodh Sansthan.

Ranawat, Manohar Singh (ed.), *Chittor–Udaipur Ka Patnama*, II, Udaipur: Chirag Prakashan, 2003.

Sagar, Agam, 'Sadri Gajal', in *Maru Bharti*, no. 4, January, Pilani: Birla Education Trust, 1981.

Sagar, Jaswant, 'Udaipur Varnan Chhand', ed. Agar Chand Nahta in *Maru Bharti*, vol. 6, pt. 4, Pilani: Birla Education Trust, January 1959.

'Sisod Vansavali Evam Rajasthan ke Rajwadon ki Vansavaliyan', in *Majjhamika*, vol. 23, Udaipur: Pratap Shodh Sansthan, 1995.

Suba-ri-Sirkar-re-Pargana-ri-Vigat (Mss), no. 227/3, Bikaner: Anup Sanskrit Library.

Thikana Bedla ka Itihas', in *Majjhamika*, no. 37, Udaipur: Pratap Shodh Sansthan, 2002.

Yati, Jai Chand, *Saiki*, ed. Muni Kanti Sagar, Jaipur: Chhutan Lal Bairathi, n.d.

INSCRIPTIONS

'Inscription of AD 120', in *Annual Report Rajputana Museums*, Ajmer: Rajputana Museum, 1931.

'Mehta, R.N., 'Survey of the Sudarshan Lake', in *Journal of Oriental Institute*, XIII (I and 2).

'Menal Inscription', in *Progress Report of Archaeological Survey*, Western Circle, 1905–6.

Vyas, A.K., 'First and Third Slabs of Kumbhalgarh Inscription VS 1517', in *Epigraphia Indica*, vol. XXIV, 1937–8, ed. N.P. Chakravarti, Delhi: Department of Archaeology, 1942.

———, 'Kadmal Plates of Guhila Vijay Sinha, VS 1140', in *Epigraphia Indica*, vol. XXXI, 1955–6, ed. D.C. Sircar, Delhi: Department of Archaeology, Managar of Publications, 1960.

———, 'Bijoli Rock Inscription of Chahmana Samesvara', in *Epigraphia Indica*, vol. XXVI, Calcutta: Department of Archaeology, 1952.

GAZETTEERS, DICTIONARIES, ATLAS AND REFERENCE BOOKS

Census of India, 1881, 1891, 1901, 1911, 1921, 1931, 1941, and 1951.

Erskine, K.D., *Rajputana Gazetteers: The Mewar Residency*, vol. I, Gurgaon: Vintage Books, rpt., 1992.

Garrick, H.B.W. and A. Cunningham, *Report of a Tour in the Punjab and Rajputana*, vol. XXIII, Delhi: Indological Book House, rpt., n.d.

Grierson, G.A., *Linguistic Survey of India*, vol. IX, pt. II, Delhi: Low Price Publications, rpt. 1990.

Lalas, Sita Ram, *Rajasthani Sabad Kos*, I, Jodhpur: Rajasthani Shodh Sansthan, 1962.

National Atlas of India, Population Jaipur Plate 113, Survey of India Maps and Census of India 1951, Government of India, 1960.

Sakariya, B.P. and Bhupatiram Skariya, *Rajasthani Hindi Sabad Kos*, Jaipur: Panchsheel Prakashan, 3 vols.: 1 (1977), 2 (1983), 3 (1984).

Wilson, H.H., *Glossary of Judicial and Revenue Terms*, Delhi: Munshiram Manoharlal, 2nd edn., 1968.

SECONDARY WORKS

Agarwal, Anil and Sunita Narain (eds.), *Dying Wisdom: Rise, Fall and Potential of India's Traditional Water Harvesting Systems*, Delhi: Centre for Science and Environment, 1997.

Bhadani, B.L., *Peasants, Artisans and Entrepreneurs: Economy of Marwar in the Seventeenth Century*, Jaipur: Rawat Publications, 1999.

Chattopadhyaya, B.D., *The Making of Early Medieval India*, Delhi: Oxford University Press, 1994.

Das, Shyamal, *Vir Vinod*, vols. I and II, pt. I, Delhi: B.R. Publishing Corporation, rpt., 1986.

Forbes, James, *Oriental Memoirs: A Narrative of Seventeen Years Residence in India*, Delhi: Gyan Publishing House, rpt., 1988.

Habib, Irfan, *An Atlas of the Mughal Empire*, Delhi: Oxford University Press, 1982.

———, *The Agrarian System of Mughal India*, Bombay: Asia Publishing House, 1963 and rev. edn., Delhi: Oxford University Press, 1999.

Hill, Donald, *A History of Engineering in Classical and Medieval Times*, London: Croom Helm, 1984.

Jain, K.C., *Ancient Cities and Towns of Rajasthan: A Study of Culture and Civilization*, Delhi: Motilal Banarsidass, 1972.

Kosambi, D.D., *Introduction to the Study of Indian History*, rev. 2nd edn., Mumbai: Popular Prakashan, 1975.

Mishra, Anupam, *Aaj Bhi Khare Hain Talab*, Delhi: Gandhi Shanti Pratisthan, 1994.

———, *Rajasthan ki Rajat Bundeyen*, Delhi: Gandhi Shanti Pratisthan, 1995.

Neubauer, Jutta Jain, *The Stepwells of Gujarat in Art-Historical Perspective*, Delhi: Abhinav Publications, 1981.

Ojha, G.H., *Udaipur Rajya ka Itihas*, vol. I, Jodhpur: Rajasthani Granthagar, rpt., 1999.

Ranawat, Ishwar Singh, *Rajasthan ke Jal Sansadhan*, Udaipur: Chirag Prakashan, 2004.

Sharma, G.N., *Social Life in Medieval Rajasthan*, Agra: L.N. Agarwal, 1968.

Sharma, R.C., *Settlement Geography of the Desert*, Delhi: Kumar Brothers, 1972.

Sinha Kapur, Nandini, *State Formation in Rajasthan*, Delhi: Manohar, 2002.

Somani, Ram Vallabh, *History of Mewar (from Earliest Times to 1751 AD)*, Jaipur: C.L. Ranka and Co., 1976.

Tod, James, *Annals and Antiquities of Rajasthan*, I, London: Routledge and Kegan Paul Ltd., rpt., 1960.

Vyas, Rajshekhar, *Mewar ki Kala aur Sthapatya*, Jaipur: Rajasthan Prakashan, 1988.

Yazdani, Gulam, *Bidar: Its History and Monuments*, London: Oxford University Press, 1947.

ARTICLES

Bhadani, B.L., 'Udai Sagar: A Sixteenth-Century Dam in Rajasthan' *Medieval India 2*, ed. Shahabuddin Iraqi, Delhi: Manohar, 2008.

Chattopadhyaya, B.D., 'Irrigation in Medieval Rajasthan', *Journal of the Economic and Social History of the Orient*, vol. XVI, pt. II, III, Leiden, 1973.

Habib, Irfan, 'Technology Changes and Society, 13th and 14th Centuries' (Presidential Address), Varanasi: *PIHC*, 1969.

Hardiman, David, 'Well Irrigation in Gujarat: Systems of Use Heirarchies of Control', in *Economic and Political Weekly*, vol. XXXIII, 20–6 June, Mumbai: Sameeksha Trust Publications, 1998.

Khan, Iqtidar Alam and Ravindra Kumar, 'The Mansagar Dam of Amber', in Anirudh Ray and S.K. Bagchi (eds.), *Technology in Ancient and Medieval India*, Delhi: Sundeep, 1986.

Khan, Mohammed Afzal 'Urban and Palace Water-Works During the Mughal Period', *Puratatva*, no. 9, 1988–9, Delhi: Indian Archaeological Society.

Lyons, Tryna, 'The Changing Faces of Eklingji: A Dynastic Shrine and Its Artists', *Artibus Asiae*, vol. LVIII, 3/4, Washington, D.C.: Smithsonian Institution.

Ojha, Jamnesh Kumar, 'Kanor Jagir ka Patta', in *Shodh-Sadhna*, no. 4, Sitamau: Natnagar Shodh Sansthan, 1983.

Qaisar, A.J., 'Agricultural Technology Depicted in Mughal Paintings', in *Itinerario*, vol. XVI, no. 2, 1992, Leiden, Grafaria, 1992.

Rosin, R. Thomas, 'Locality and Frontier Securing Livelihood in the Aravalli Zone of Central Rajasthan', in Karine Schomer, Joan L. Erdman, Deryck O. Lodrick and Lloyd I. Rudolph, eds., *The Idea of Rajasthan*, II, Delhi: Manohar, 2001.

Sharma, Shakti Kumar, 'Shakunt', 'Jai Singh Prashasti Main Jai Samudra: Nirman ka Vivran', in K.L. Sharma, ed., *Rajasthan ka Aitihasik Sanskrit Sahitya*, Jodhpur: RORI, 1996.

Sharma, Rajiv, 'Raj Samand Dam: An Achievement of the 17th Century Civil Engineering', *Proceedings of Indian History Congress*, 1992–3.

Singh, Vinod Kumar, 'A Chandella Reservoir: Kalyan Sagar of Mahoba', *Proceedings of Indian History Congress*, 53rd Session, Warangal, 1992–3.

Vashishtha, Neelima, 'A Note on the Sculptors of the Raj Samudra Lake in Rajasthan', in *Art and Architecture* (felicitation volume in Honour of Professor S. Nurul Hasan), ed. Ahsan Jan Qaisar and Som Prakash Verma, Jaipur: Publication Scheme, 1993.

Index